FPGAs
Fundamentals, Advanced Features, and Applications in Industrial Electronics

FPGAs
Fundamentals, Advanced Features, and Applications in Industrial Electronics

Juan José Rodríguez Andina,
Eduardo de la Torre Arnanz, and
María Dolores Valdés Peña

CRC Press
Taylor & Francis Group
Boca Raton London New York

CRC Press is an imprint of the
Taylor & Francis Group, an **informa** business

CRC Press
Taylor & Francis Group
6000 Broken Sound Parkway NW, Suite 300
Boca Raton, FL 33487-2742

© 2017 by Taylor & Francis Group, LLC
CRC Press is an imprint of Taylor & Francis Group, an Informa business

No claim to original U.S. Government works

Printed on acid-free paper
Version Date: 20161025

International Standard Book Number-13: 978-1-4398-9699-0 (Hardback)

Visit the Taylor & Francis Web site at
http://www.taylorandfrancis.com

and the CRC Press Web site at
http://www.crcpress.com

Contents

Preface

This book intends to contribute to a wider use of field-programmable gate arrays (FPGAs) in industry by presenting the concepts associated with this technology in a way accessible for nonspecialists in hardware design so that they can analyze if and when these devices are the best (or at least a possible) solution to efficiently address the needs of their target industrial applications. This is not a trivial issue because of the many different (but related) factors involved in the selection of the most suitable hardware platform to solve a specific digital design problem. The possibilities enabled by current FPGA devices are highlighted, with particular emphasis on the combination of traditional FPGA architectures and powerful embedded processors, resulting in the so-called field-programmable systems-on-chip (FPSoCs) or systems-on-programmable-chip (SoPCs). Discussions and analyses are focused on the context of embedded systems, but they are also valid and can be easily extrapolated to other areas.

The book is structured into nine chapters:

- Chapter 1 analyzes the different existing design approaches for embedded systems, putting FPGA-based design in perspective with its direct competitors in the field. In addition, the basic concept of FPGA "programmability" or "configurability" is discussed, and the main elements of FPGA architectures are introduced.

- From the brief presentation in Chapter 1, Chapter 2 describes in detail the main characteristics, structure, and *generic* hardware resources of modern FPGAs (logic blocks, I/O blocks, and interconnection resources). Some specialized hardware blocks (clock management blocks, memory blocks, hard memory controllers, transceivers, and serial communication interfaces) are also analyzed in this chapter.

- Embedded soft and hard processors are analyzed in Chapter 3, because of their special significance and the design paradigm shift they caused as they transformed FPGAs from hardware accelerators to FPSoC platforms. As shown in this chapter, devices have evolved from simple ones, including one general-purpose microcontroller, to the most recent ones, which integrate several (more than 10 in some cases) complex processor cores operating concurrently, opening the door for the implementation of homogeneous or heterogeneous multicore architectures. The efficient communication between processors and their peripherals is a key factor to successfully develop embedded systems. Because of this, the currently available on-chip buses and their historical evolution are also analyzed in detail in this chapter.

- Chapter 4 analyzes DSP blocks, which are very useful hardware resources in many industrial applications, enabling the efficient implementation of key functional elements, such as digital filters, encoders, decoders, or mathematical transforms. The advantages provided by the inherent parallelism of FPGAs and the ability of most current devices to implement floating-point operations in hardware are also highlighted in this chapter.

- Analog blocks, including embedded ADCs and DACs, are addressed in Chapter 5. They allow the functionality of the (mostly digital) FPGA devices to be extended to simplify interfacing with the analog world, which is a fundamental requirement for many industrial applications.

- The increasing complexity of FPGAs, which is clearly apparent from the analyses in Chapters 2 through 5, can only be efficiently handled with the help of suitable software tools, which allow complex design projects to be completed within reasonably short time frames. Tools and methodologies for FPGA design are presented in Chapter 6, including tools based on the traditional RTL design flow, tools for SoPC design, high-level synthesis tools, and tools targeting multithread accelerators for high-performance computing, as well as debugging and other auxiliary tools.

- There are many current applications where tremendous amounts of data have to be processed. In these cases, communication resources are key elements to obtain systems with the desired (increasingly high) performance. Because of the many functionalities that can be implemented in FPGAs, such efficient communications are required to interact not only with external elements but also with internal blocks to exchange data at the required rates. The issues related to both off-chip and in-chip communications are analyzed in detail in Chapter 7.

- The ability to be reconfigured is a very interesting asset of FPGAs, which resulted in a new paradigm in digital design, allowing the same device to be readily adapted during its operation to provide different hardware functionalities. Chapter 8 focuses on the main concepts related to FPGA reconfigurability, the advantages of using reconfiguration concurrently with normal operation (i.e., at run time), the different reconfiguration alternatives, and some existing practical examples showing high levels of hardware adaptability by means of run-time dynamic and partial reconfiguration.

- Today, FPGAs are used in many industrial applications because of their high speed and flexibility, inherent parallelism, good cost–performance trade-off (offered through wide portfolios of different device families), and the huge variety of available specialized logic resources. They are expected not only to consolidate their application domains but also to enter new ones. To conclude the book, Chapter 9 addresses industrial applications of FPGAs in three main design areas (advanced control techniques, electronic instrumentation, and digital real-time simulation) and three very significant application domains (mechatronics, robotics, and power systems design).

Acknowledgments

The authors have greatly benefited during their more than 25 years of experience in FPGA design from advice and comments from, and discussions with, many colleagues, from both academia and the industry. Citing all of them individually here is not possible and might result in some unintentional omission. We hope all of them know they are represented here through our grateful acknowledgments to our present and past colleagues and students at the Department of Electronic Technology, University of Vigo, and the Center of Industrial Electronics, Technical University of Madrid; the people at the many companies for which we have consulted and developed projects in the area; our colleagues in the IEEE Industrial Electronics Society; and those we have met over the years in many scientific forums, such as IECON, ISIE, ICIT, FPL, Reconfig, and ReCoSoc.

Last, but of course not the least, our final word of gratitude goes to our families for their unconditional support.

Authors

Juan José Rodríguez Andina received his MSc from the Technical University of Madrid, Spain, in 1990, and his PhD from the University of Vigo, Spain, in 1996, both in electrical engineering. He has also received the Extraordinary Doctoral Award from the University of Vigo. He is an associate professor in the Department of Electronic Technology, University of Vigo. In 2010–2011, he was on sabbatical as a visiting professor at the Advanced Diagnosis, Automation, and Control Laboratory, Electrical and Computer Engineering Department, North Carolina State University, Raleigh. He has been working for more than 25 years in digital systems design, with emphasis on FPGA-based design for industrial applications. He has authored more than 140 journal and conference articles and holds several Spanish, European, and U.S. patents. He currently serves as vice president for conference activities of the IEEE Industrial Electronics Society and has been general chair, technical program chair, and member of other various committees in a number of IEEE conferences (such as IECON, ISIE, ICIT, and INDIN), where he regularly organizes special sessions related to industrial applications of FPGAs and embedded systems. He is the former editor-in-chief of the *IEEE Industrial Electronics Magazine* and an associate editor for *IEEE Transactions on Industrial Electronics* and *IEEE Transactions on Industrial Informatics*.

Eduardo de la Torre Arnanz is an associate professor of electronics since 2002 and obtained his MSc and PhD in electrical engineering from the Technical University of Madrid in 1989 and 2000, respectively. His main expertise is in FPGA-based design and, in particular, in partial and dynamic reconfiguration of digital systems and reconfigurable hardware acceleration. He has been working for more than 25 years on digital systems design, among which more than 20 have been around FPGAs, mostly in industrial applications. He has authored more than 40 papers on reconfigurable systems in the last five years and has been program cochair of ReCoSoC (2015), Reconfig (2012 and 2013), DASIP (2013), and SPIE VLSI Circuits & Systems (2009 and 2011) conferences as well as a program committee member of conferences such as FPL, ReCoSoC, RAW, WRC, ISVLSI, and SIES. He is also a reviewer of numerous conferences and journals such as the *IEEE Transactions on Computers, IEEE Transactions on Industrial Informatics, IEEE Transactions on Industrial Electronics*, and *Sensor* magazine.

 **María Dolores Valdés Peña** is an associate professor in the Department of Electronic Technology, University of Vigo, Spain. She received her MSc from Universidad Central de Las Villas, Santa Clara, Cuba, in 1990, and her PhD from the University of Vigo, Vigo, Spain, in 1997, both in electrical engineering. She received the Extraordinary Doctoral Award from the University of Vigo. In 1998, the Society of Instrument and Control Engineers (SICE) of Japan gave her the award for the best research work at the 37th Annual SICE Conference. Over the years, she has authored more than 120 journal and conference articles. Her research interests include the design of reconfigurable systems based on FPGAs applied to data acquisition and conditioning systems, digital signal processing and control, wireless sensor networks, and field-programmable systems-on-chip for industrial applications.

1

FPGAs and Their Role in the Design of Electronic Systems

1.1 Introduction

This book is mainly intended for those users who have had certain experience in digital control systems design, but for some reason have not had the opportunity or the need to design with modern field-programmable gate arrays (FPGAs). The book aims at providing a description of the possibilities of this technology, the methods and procedures that need to be followed in order to design and implement FPGA-based systems, and selection criteria on what are the best suitable and cost-effective solutions for a given problem or application. The focus of this book is on the context of embedded systems for industrial use, although many concepts and explanations could be also valid for other fields such as high-performance computing (HPC). Even so, the field is so vast that the whole spectrum of specific applications and application domains is still tremendously large: transportation (including automotive, avionics, railways systems, naval industry, and any other transportation systems), manufacturing (control of production plants, automated manufacturing systems, etc.), consumer electronics (from small devices such as an air-conditioning remote controller to more sophisticated smart appliances), some areas within the telecom market, data management (including big data), military industry, and so forth.

In this chapter, the concept of embedded systems is presented from a wide perspective, to later show the ways of approaching the design of embedded systems with different complexities. After introducing all possibilities, the focus is put on FPGA-related applications. Then, the basic concept of FPGA "programmability" or "configurability" is discussed, going into some description of the architectures, methods, and supported tools required to successfully carry out FPGA designs with different complexities (not only in terms of size but also in terms of internal features and design approaches).

1.2 Embedded Control Systems: A Wide Concept

Embedded control systems are, from a very general perspective, control elements that, in a somewhat autonomous manner, interact with a physical system in order to have an automated control over it. The term "embedded" refers to the fact that they are placed in or nearby the physical system under control. Generally speaking, the interfaces between the physical and control systems consist of a set of sensors, which provide information from the physical system to the embedded system, and a set of actuators capable, in general, of modifying the behavior of the physical system.

Since most embedded systems are based on digital components, signals obtained from analog sensors must be transformed into equivalent digital magnitudes by means of the corresponding analog-to-digital converters (ADCs). Equivalently, analog actuators are managed from digital-to-analog converters (DACs). In contrast, digital signals do not require such modifications. The success of smart sensor and actuator technologies allows such interfaces to be simplified, providing standardized communication buses as the interface between sensors/actuators and the core of the embedded control system.

Without loss of generality regarding the earlier paragraphs, two particular cases are worth mentioning: communication and human interfaces. Although both would probably fit in the previously listed categories, their purposes and nature are quite specific.

On one hand, communication interfaces allow an embedded system to be connected to other embedded systems or to computing elements, building up larger and more complex systems or infrastructures consisting of smaller interdependent physical subsystems, each one locally controlled by their own embedded subsystem (think, for instance, of a car or a manufacturing plant with lots of separate, but interconnected, subsystems).

Communication interfaces are "natural" interfaces for embedded control systems since, in addition to their standardization, they take advantage from the distributed control system philosophy, providing scalability, modularity, and enhanced dependability—in terms of maintainability, fault tolerance, and availability as defined by Laprie (1985).

On the other hand, human interfaces can be considered either like conventional sensors/actuators (in case they are simple elements such as buttons, switches, or LEDs) or like simplified communication interfaces (in case they are elements such as serial links for connecting portable maintenance terminals or integrated in the global communication infrastructure in order to provide remote access). For instance, remote operation from users can be provided by a TCP/IP socket using either specific or standard protocols (like http for web access), which easily allows remote control to be performed from a web browser or a custom client application, the server being embedded in the control system. Nowadays, nobody gets surprised by the possibility of

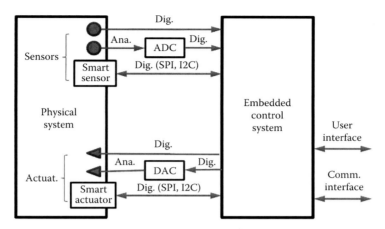

FIGURE 1.1
Generic block diagram of an embedded system.

using a web browser to access the control of a printer, a photocopy machine, a home router, or a webcam in a ski resort.

Figure 1.1 presents a general diagram of an embedded control system and its interaction with the physical system under control and other subsystems.

Systems based on analog sensors and actuators require signal conditioning operations, such as low-noise amplification, anti-aliasing filtering, or filtering for noise removal, to be applied to analog signals. Digital signal processing and computationally demanding operations are also usually required in this case. On the other hand, discrete sensors and actuators tend to make the embedded system more control dependent. Since they have to reflect states of the system, complexity in this case comes from the management of all state changes for all external events. As a matter of fact, medium- or large-size embedded systems usually require both types of sensors and actuators to be used. On top of that, in complex systems, different control subtasks have to be performed concurrently since the key to achieve successful designs is to apply the "divide and conquer" approach to the global system, in order to break down its functionality into smaller, simpler subsystems.

As one might think, the previous paragraphs may serve as introductory section for a book on any type of embedded systems, these being based on microcontrollers, computers, application-specific integrated circuits (ASICs), or (of course) FPGAs. Therefore, since implementation platforms do not actually modify the earlier definitions and discussion significantly, one of the main objectives of this book is to show when and how FPGAs could (or should) be used for the efficient implementation of embedded control systems targeting industrial applications. Since each technology has its own advantages and limitations, decision criteria must be defined to select the technology or technologies best suited to solve a given problem. Fairly speaking, the authors do not claim FPGAs to be used for any industrial control system, but their intention is to help designers identify the cases where FPGA technology provides advantages (or is the only possibility) for the implementation of embedded systems in a particular application or application domain.

1.3 Implementation Options for Embedded Systems

Selecting the most suitable technique to implement an embedded system that fulfills all the requirements of a given application may not be a trivial issue since designers need to consider many different interrelated factors. Among the most important ones are cost, performance, energy consumption, available resources (i.e., computing resources, sizes of different types of memories, or the number and type of inputs and outputs available), reliability and fault tolerance, or availability. Even if these are most likely the factors with higher impact on design decision, many others may also be significant in certain applications: I/O signal compatibility, noise immunity (which is strongly application dependent), harsh environmental operating conditions (such as extreme temperature or humidity), tolerance to radiations, physical size restrictions, special packaging needs, availability of the main computing device and/or of companion devices (specific power supplies, external crystal oscillators, specific interfaces, etc.), existence of second sources of manufacturing, time to product deprecation, intellectual property (IP) protection, and so forth.

For simple embedded systems, small microcontrollers and small FPGAs are the main market players. As the complexity of the applications to be supported by the embedded system grows, larger FPGAs have different opponents, such as digital signal processing (DSP) processors, multicore processors, general-purpose graphic processing units (GPGPUs), and ASICs. In order to place the benefits and drawbacks of FPGAs within this contest, qualitative and quantitative comparisons between all these technologies are presented in the next sections for readers to have sound decision criteria to help determine what are the most appropriate technologies and solutions for a given application.

1.3.1 Technological Improvements and Complexity Growth

The continuous improvements in silicon semiconductor fabrication technologies (mainly resulting in reductions of both transistor size and power supply voltage) implicitly allow lower energy consumption and higher performance to be achieved. Transistor size reduction also opens the door for higher integration levels, which although undoubtedly being a big advantage for designers, give rise to some serious threats regarding, for instance, circuit reliability, manufacturing yield (i.e., the percentage of fabricated parts that work correctly), or noise immunity.

Power consumption is also becoming one of the main problems faced by designers, not only because of consumption itself but also because of the need for dissipating the resulting heat produced in silicon (especially with modern 3D stacking technologies, where different silicon dies are decked, reducing the dissipation area while increasing the number of transistors—and, therefore, the power consumption—per volume unit). Circuits at the edge of the technology are rapidly approaching the limits in this regard, which are estimated to be around 90 W/cm^2, according to the challenges reported for reconfigurable computing by the High Performance and Embedded Architectures and Computation Network of Excellence (www.hipeac.net).

The integration capacity is at the risk of Moore's law starting to suffer from some fatigue. As a consequence, the continuous growth of resource integration over the years is slowing down compared to what has been happening over the last few decades. Transistor sizes are not being reduced at the same pace as higher computing performance is being demanded, so larger circuits are required. Larger circuits negatively affect manufacturing yield and are negatively affected by process variation (e.g., causing more differences to exist between the fastest and slowest circuits coming out from the same manufacturing run, or even having different parts of the same circuit achieving different maximum operating frequencies). This fact, combined with the use of lower power supply voltages, also decreases fault tolerance, which therefore needs to be mitigated by using complex design techniques and contributes to a reduction in system lifetime.

Maximum operating frequency seems to be saturated in practice. A limit of a few GHz for clock frequencies has been reached in regular CMOS technologies with the smallest transistor sizes, no matter the efforts of circuit designers to produce faster circuits, for instance, by heavily pipelining their designs to reduce propagation delay times of logic signals between flip-flops (the design factor that, apart from the technology itself, limits operating frequency).

Is the coming situation that critical? Probably not. These problems have been anticipated by experts in industry and academia, and different approaches are emerging to handle most of the issues of concern. For instance, low-power design techniques are reaching limits that were not imaginable 10–15 years ago, using dynamic voltage scaling or power gating and taking advantage of enhancements in transistor threshold voltages (e.g., thanks to multithreshold transistors). Anyway, the demand for higher computing power with less energy consumption is still there. Mobile devices have a tremendous, ever-increasing penetration in all aspects of our daily lives, and the push of all these systems is much higher than what technology, alone, can handle.

Is there any possibility to face these challenges? Yes, using parallelism. Computer architectures based on single-core processors are no longer providing better performance. Different smart uses of parallel processing are leading the main trends in HPC, as can be seen in the discussion by Kaeli and Akodes (2011). Actually, strictly speaking, taking the most advantage of parallelism does not just mean achieving the highest possible performance using almost unlimited computing resources but also achieving the best possible performance–resources and performance–energy trade-offs. This is the goal in the area of embedded systems, where resources and the energy budget are limited.

In this context, hardware-based systems and, in particular, configurable devices are opening new ways for designing efficient embedded systems. Efficiency may be roughly measured by the ratio of the *number of operations per unit of energy consumed*, for example, in MFlops/mW (millions of floating-point operations per second per milliwatt). Many experiments have shown that improvements of two orders of magnitude may be achieved by replacing single processors with hardware computing.

1.3.2 Toward Energy-Efficient Improved Computing Performance

FPGAs have an increasingly significant role when dealing with energy consumption. Sometime ago, the discussion regarding consumption would have been centered on *power* consumption, but the shift toward considering also *energy* as a major key element comes from the fact that the concern on energy availability is becoming a global issue. Initially, it just affected portable devices or, in a more general sense, battery-operated devices with limited usability because of the need of recharging or replacing batteries.

However, the issue of energy usage in any computing system is much more widespread. The most opposite case to restricted-energy, restricted-resource tiny computing devices might be that of huge supercomputing centers (such as for cloud-computing service providers). The concern on the "electricity bill" of these companies is higher than ever. To this respect, although FPGAs cannot be considered the key players in this area, they are presently having a growing penetration. As a proof of that, it can be noticed that there are services (including some any of us could be using daily, such as web searchers or social network applications) currently being served by systems including thousands of FPGAs instead of thousands of microprocessors.

Why is this happening? Things have changed in recent years, as technologies and classical computing architectures are considered to be mature enough. There are some facts (slowly) triggering a shift from software-based to hardware-based computing. As discussed in Section 1.3.1, fabrication technologies are limited in the maximum achievable clock speed, so no more computing performance can be obtained from this side. Also, single-core microprocessor architectures have limited room for enhancement. Even considering complex cache or deep pipelined structures, longer data size operators, or other advanced features one might think of, there are not much chances for significant improvements.

There is no way of significantly improving performance in a computing system other than achieving higher levels of parallelism. To this respect, the trend in software-based computing is to move from single-core architectures to multi- or many-core approaches. By just taking into account that hardware-based computing structures (and, more specifically, FPGAs) are intrinsically parallel, it means that FPGAs have better chances than software-based computing solutions to provide improved computing performance (Jones et al. 2010).

In addition, if the energy issue is included in the equation, it is clear that reducing the time required to perform a given computation helps in reducing the total energy consumption associated with it (moreover, considering that, in most applications, dynamic power consumption is dominant over static power consumption*). Thus, a certain device with higher power consumption than an opponent device may require less energy to complete a computing task if the acceleration it provides with respect to the opponent is sufficiently high. This is the case with FPGAs, where the computing fabric can be tailored to exploit the best achievable parallelism for a given task (or set of tasks), having in mind both acceleration and energy consumption features.

1.3.3 A Battle for the Target Technology?

The graph in Figure 1.2 qualitatively shows the performance and flexibility offered by the different software- and hardware-based architectures suitable for embedded system design. Flexibility is somewhat related to the ease of use of a system and its possibilities to be adapted to changes in specifications.

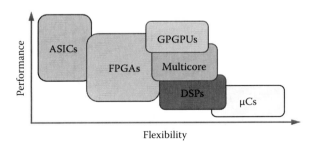

FIGURE 1.2
Performance versus flexibility of different technologies for embedded system design.

---

* Even though, since technologies with higher integration levels and higher resource availability suffer more from static power consumption, this factor needs to be taken into consideration.

As can be seen from Figure 1.2, the highest performance is achieved by ASICs and GPGPUs. The lack of flexibility of ASICs is compensated by their very high energy efficiency. In contrast, although GPGPUs are excellent (if not the best) performant software-based computing devices, they are highly power consuming. Multicore technologies are close to GPGPUs in performance, and, in some cases, their inherent parallelism matches better the one required by the target application. While GPGPUs exploit data parallelism more efficiently, multicore systems are best suited to multitask parallelism. Known drawbacks of GPGPUs and many multicore systems include the need for relying on a host system and limited flexibility with respect to I/O availability. At the high end of flexibility, DSP processors offer better performance than single general-purpose microprocessors or microcontrollers because of their specialization in signal processing. This, of course, is closely related to the amount of signal processing required by the target application.

FPGAs are represented as less flexible than specialized and general-purpose processors. This comes from the fact that software-based solutions are in principle more flexible than hardware-based ones. However, there exist FPGAs that can be reconfigured during operation, which may ideally be regarded to be as flexible as software-based platforms since reconfiguring an FPGA essentially means writing new values in its configuration memory. This is very similar to modifying a program in software approaches, which essentially means writing the new code in program memory. Therefore, FPGAs can be considered in between ASICs and software-based systems, in the sense that they have hardware-equivalent speeds with software-equivalent flexibility. Shuai et al. (2008) provide a good discussion on FPGAs and GPUs.

Figure 1.3 shows a comparative diagram of the aforementioned approaches. The legend shows the axes for flexibility, power efficiency, performance, unit cost, and design cost (complexity).

The cost associated with design complexity is important for embedded devices because the number of systems to be produced may not be too large. Since in the area of embedded systems, significant customization and design effort are needed for every design, additional knowledge is demanded as complexity grows. Hence, complex systems might require a lot of design expertise, which is not always available in small- or medium-sized design offices and labs. Design techniques and tools are therefore very important in embedded system design. They are briefly analyzed in Section 1.3.4. Moreover, tools related to FPGA-based design are analyzed in detail in Chapter 6.

1.3.4 Design Techniques and Tools for the Different Technologies

Some design techniques and tools (e.g., those related to PCB design and manufacturing) are of general applicability to all technologies mentioned so far. According to the complexity of the design, these might include techniques and tools for electromagnetic protection and emission mitigation,

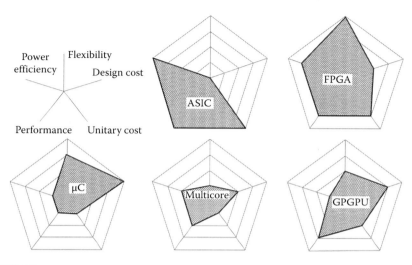

FIGURE 1.3
Comparative features of ASICs, FPGAs, general-purpose processors and microcontrollers, multicore processors, and GPGPUs. *Notes:* Outer (further away from the center) is better. ASIC cost applies to large enough production.

thermal analysis, signal integrity tests, and simulation. Some other techniques and tools are specific to the particular technology used (or to some but not all of them), as discussed in the following sections.

1.3.4.1 General-Purpose Processors and Microcontrollers

General-purpose processors and microcontrollers are the best (and sometimes the only practically viable) solution for simple embedded systems with low computing power requirements. They just require designers' knowledge in programming and debugging simple programs, mostly (if not totally) written in high-level languages. Emulators and debuggers help in validating the developed programs, contributing to a fast design cycle.

In the simplest cases, no operating system (OS) is required; the processor just runs the application program(s), resulting in the so-called bare-metal system. As complexity grows, applications might require complex use of interrupts or other processor features, therefore making it necessary to use an OS as a supporting layer for running multiple tasks concurrently. In this case, dead times of a task can be used to run other tasks, giving some impression of parallelism (although actually it is just concurrency). In very specific cases, system requirements might demand the use of assembly code in order to accelerate some critical functions.

In brief, off-the-shelf simple and cheap processors must be used whenever they are powerful enough to comply with the requirements of simple target applications. Other platforms should be considered only if they provide a significant added value. For instance, it may be worth using FPGAs to solve a simple application if, in addition, they implement glue logic and avoid the need for populating PCBs with annoying discrete devices required just for interconnection or interfacing purposes.

1.3.4.2 DSP Processors

Because of their specific architectures, DSP processors are more suitable than general-purpose processors or microcontrollers for many applications above a certain complexity level, where they provide better overall characteristics when jointly considering performance, cost, and power consumption. For instance, DSP-based platforms can solve some problems by working at lower clock frequencies than general-purpose processors would require (therefore reducing energy consumption) or achieving higher throughput (if they work at the same clock frequency).

Specific DSP features are intended to increase program execution efficiency. These include hardware-controlled loops, specialized addressing modes (e.g., bit reverse, which dramatically reduces execution times in fast Fourier transforms), multiply–accumulate (MAC) units (widely used in many signal processing algorithms), multiple memory banks for parallel access to data, direct memory access (DMA) schemes for automated fast I/O, and the ability to execute some instructions in a single clock cycle. There are complex issues (both hardware and software) related to the efficient use of these features. For instance, real parallel operation of functional units can be achieved by using very long instruction word (VLIW) architectures. VLIW systems are capable of determining (at compile time) whether several functional units can be used simultaneously (i.e., if at a given point of program execution there are instructions requiring different functional units to carry out operations with different data that are already available). This implies the ability of the hardware to simultaneously access several memories and fetch several operands or coefficients, which are simultaneously sent to several functional units. In this way, by using multiple MAC units, it may be possible, for instance, to compute several stages of a filter in just one clock cycle.

Advantage can only be taken from DSP processors with VLIW architectures with deep knowledge of the architecture and carefully developing assembly code for critical tasks. Otherwise, they may be underused, and performance could even be worse than when using more standard architectures.

For those embedded systems where the performance of off-the-shelf DSP processors complies with the requirements of the application and that of general-purpose processors or microcontrollers does not, the former are in general the best implementation platform. However, like in the case of general-purpose processors and microcontrollers, except for some devices tailored for specific applications, it is not unusual that external acquisition circuitry is required (e.g., for adaptation to specific I/O needs), which may justify the use of FPGAs instead of DSP processors.

Although DSP processors exploit parallelism at functional module level, it might be the case that the maximum performance they offer is not enough for a given application. In this case, real parallel platforms need to be used. Software-based parallel solutions (multicore processors and GPGPUs) are discussed in Section 1.3.4.3 and hardware-based ones (FPGAs) in Section 1.3.4.4.

1.3.4.3 Multicore Processors and GPGPUs

The internal architectures of multicore processors and GPGPUs are designed to match task parallelism and data parallelism, respectively. Multicore systems can very efficiently execute multiple, relatively independent tasks, which are distributed among a network of processing cores, each of them solving either a different task or some concurrent tasks. GPGPUs contain a large number of computing elements executing threads that, in a simplistic manner, may be considered as relatively (but not fully) independent executions of the same code over different pieces of data.

Multicore devices can be programmed using conventional high-level languages (such as C or C++), just taking into consideration that different portions of the code (i.e., different tasks) are to be assigned to different processors. The main issues regarding the design with these platforms are related to the need for synchronization and data transfer among tasks, which are usually addressed by using techniques such as semaphores or barriers, when the cores share the same memory, or with message passing through interconnection networks, when each core has its own memory. These techniques are quite complex to implement (in particular for shared memory systems) and also require detailed, complex debugging. For systems with hard real-time constraints, ensuring the execution of multiple tasks within the target deadlines becomes very challenging, although some networking topologies and resource management techniques can help in addressing, to a certain extent, the predictability problem (not easily, though).

GPGPUs operate as accelerators of (multi)processor cores (hosts). The host runs a sequential program that, at some point in the execution process, launches a kernel consisting of multiple threads to be run in the companion GPGPU. These threads are organized in groups, such that all threads within the same group share data among them, whereas groups are considered to be independent from each other. This allows groups to be executed in parallel according to the execution units available within the GPGPU, resulting in the so-called virtual scalability. Thread grouping is not trivial, and the success of a heavily accelerated algorithm depends on groups efficiently performing memory accesses. Otherwise, kernels may execute correctly but with low performance gains (or even performance degradation) because of the time spent in data transfers to/from GPGPUs from/to hosts. Programming kernels requires the use of languages with explicit parallelism, such as CUDA or OpenCL. Debugging is particularly critical (and nontrivial) since a careful and detailed analysis is required to prevent malfunctions caused by desynchronization of threads, wrong memory coalescence policies, or inefficient kernel mapping. Therefore, in order for GPGPUs to provide better performance than the previously discussed implementation platforms, designers must have deep expertise in kernel technology and its mapping in GPGPU architectures.

1.3.4.4 FPGAs

FPGAs offer the possibility of developing configurable, custom hardware that might accelerate execution while providing energy savings. In addition, thanks to the increasing scale of integration provided by fabrication technologies, they can include, as discussed in detail in Chapter 3, one or more processing cores, resulting in the so-called field-programmable systems-on-chip (FPSoCs) or systems-on-programmable chip (SoPCs).* These systems may advantageously complement and extend the characteristics of the aforementioned single- or multicore platforms with custom hardware accelerators, which allow the execution of all or some critical tasks to be optimized, both in terms of performance and energy consumption.

Powerful design tools are required to deal with the efficient integration of these custom hardware peripherals and others from off-the-shelf libraries, as well as other user-defined custom logic, with (multiple) processor cores in an FPSoC architecture. These SoPC design tools, described in Chapter 6, require designers to have good knowledge of the underlying technologies and the relationship among the different functionalities needed for the design of FPSoCs, in spite of the fact that vendors are making significant efforts for the automation and integration of all their tools in single (but very complex) environments. In this sense, in the last few years, FPGA vendors are offering solutions for multithreaded acceleration that compete with GPGPUs, thus providing tools to specify OpenCL kernels that can be mapped into FPGAs. Also, long-awaited high-level synthesis (HLS) tools now provide a method to migrate from high-level languages such as C, C++, or SystemC into hardware description languages (HDLs), such as VHDL or Verilog, which are the most widely used today by FPGA design tools.

1.3.4.5 ASICs

ASICs are custom integrated circuits (mainly nonconfigurable, in the sense explained in Section 1.4) fully tailored for a given application. A careful design using the appropriate manufacturing technology may yield excellent performance and energy efficiency, but the extremely high nonrecurrent engineering costs and the very specific and complex skills required to design them make this target technology unaffordable for low- and medium-sized productions.

---

* Since the two acronyms may be extensively found in the literature as well as in vendor-provided information, both of them are interchangeably used throughout this book.

The lack of flexibility is also a problem, since, nowadays, many embedded systems need to be capable of being adapted to very diverse applications and working environments. For instance, the ability to adapt to changing communication protocols is an important requirement in many industrial applications.

1.4 How Does Configurable Logic Work?

First of all, it is important to highlight the intrinsic difference between *programmable* and *(re)configurable* systems. The "P" in FPGA can be misleading since, although FPGAs are the most popular and widely used *configurable* circuits, it stands for *programmable*. Both kinds of systems are intended to allow users to change their functionality. However, not only in the context of this book but also according to most of the literature and the specialized jargon, *programmable* systems (processors) are those based on the execution of software, whereas *(re)configurable* systems are those whose internal hardware computing resources and interconnects are not totally configured by default. Configuration consists in choosing, configuring, and interconnecting the resources to be used. Software-based solutions typically rely on devices whose hardware processing structure is fixed, although, as discussed in Chapter 3, the configurable hardware resources of an FPGA can be used to implement a processor, which can then obviously be programmed.

The fixed structure of programmable systems is built so as to allow them to execute different sequences (software programs) of basic operations (instructions). The programming process mainly consists in choosing the right instructions and sequences for the target application. During execution, instructions are sequentially fetched from memory, then (if required) data are fetched from memory or from registers, the processing operation implied by the current instruction is computed, and the resulting data (if any) are written back to memory or registers. As can be inferred, the hardware of these systems does not provide functionality by itself, but through the instructions that build up the program being executed.

On the other hand, in configurable circuits, the structure of the hardware resources resulting from the configuration of the device determines the functionality of the system. Using different configurations, the same device may exhibit different internal functional structures and, therefore, different user-defined functionalities. The main advantage of configurable systems with regard to pure software-based solutions is that, instead of sequentially executing instructions, hardware blocks can work in a collaborative concurrent way; that is, their execution of tasks is inherently parallel.

Arranging the internal hardware resources to implement a variety of digital functions is equivalent, from a functional point of view, to manufacturing different devices for different functions, but with configurable circuits,

no further fabrication steps are required to be applied to the premanufactured off-the-shelf devices. In addition, configuration can be done at the user
premises, or even infield at the operating place of the system.

The beginning of reconfigurable devices started with programmable*
logic matrices (programmable logic array [PLA] and programmable
array logic [PAL]—whose basic structures are shown in Figure 1.4), where
the connectivity of signals was decided using arrays of programmable connections. These were originally fuses (or antifuses†), which were selectively
either burnt or left intact during configuration.

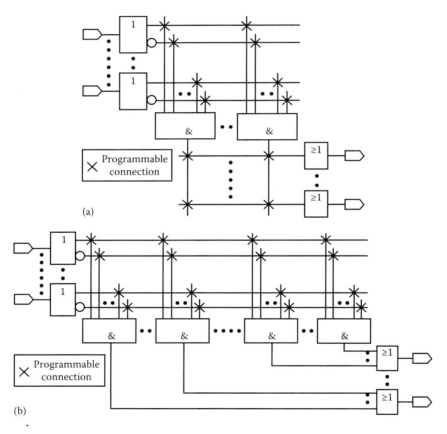

FIGURE 1.4
Programmable matrices: (a) PLA; (b) PAL.

---

* At that time, the need for differentiating programmability and configurability had not yet
been identified.
† The difference between fuses and antifuses resides in their state after being burnt, open or
short circuit, respectively.

In programmable matrices, configuration makes the appropriate input signals participate in the sums of products required to implement different logic functions. When using fuses, this was accomplished by selectively overheating those to be burnt, driving a high current through them. In this case, the structural internal modifications are literally real and final, since burnt fuses cannot be configured back to their initial state.

Although the scale of integration of fuses was in the range of several micrometers (great for those old days), CMOS integration was advancing at a much faster pace, and quite soon, new configuration infrastructures were developed in the race for larger, faster, and more flexible reconfigurable devices. Configuration is no longer based on changes in the physical structure of the devices, but on the behavior regarding connectivity and functionality, specified by the information stored in dedicated memory elements (the so-called *configuration* memory). This not only resulted in higher integration levels but also increased flexibility in the design process, since configurable devices evolved from being one-time programmable to being reconfigurable, which can be configured several times by using erasable and reprogrammable memories for configuration. Nowadays, a clear technological division can be made between devices using nonvolatile configuration memories (EEPROM and, more recently, flash) and those using volatile configuration memories (SRAM, which is the most widely used technology for FPGA configuration).

Currently, programmable matrices can be found in programmable logic devices (PLDs), which found their application niche in glue logic and finite-state machines. The basic structure of PLDs is shown in Figure 1.5. In addition to configuring the connections between rows and columns of the programmable matrices, in PLDs, it is also possible to configure the behavior of the macrocells.

The main drawback of PLDs comes from the scalability problems related to the use of programmable matrices. This issue was first addressed by including several PLDs in the same chip, giving rise to the complex PLD concept. However, it soon became apparent that this approach does not solve the scalability problem to the extent required by the ever-increasing complexity of digital systems, driven by the evolution of fabrication technologies. A change in the way configurable devices were conceived was needed. The response to that need were FPGAs, whose basic structure is briefly described in the following.*

Like all configurable devices, FPGAs are premanufactured, fixed pieces of silicon. In addition to configuration memory, they contain a large number of basic configurable elements, ideally allowing them to implement any digital system (within the limits of the available chip resources). There are two main types of building blocks in FPGAs: (relatively small) configurable logic circuits spread around the whole chip area (logic blocks [LBs]) and, between them, configurable interconnection resources (interconnect logic [IL]).

---

* FPGA architectures are analyzed in detail in Chapter 2.

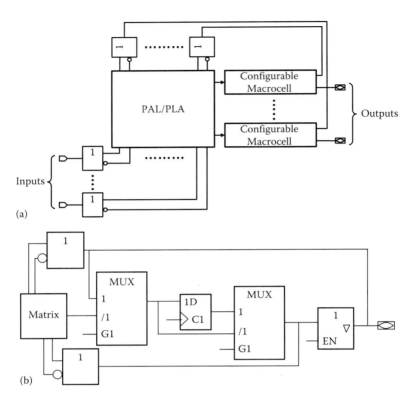

FIGURE 1.5
(a) Basic PLD structure; (b) sample basic macrocell.

The functionality of the target system is obtained by adequately configuring the behavior of the required LBs and the IL that interconnects them, by writing the corresponding information in the FPGA's internal configuration memory. The information is organized in the form of a stream of binary data (called bitstream) coming out from the design process, which determines the behavior of every LB and every interconnection inside the device. FPGA configuration issues are analyzed in Chapter 6.

A most basic LB would consist of the following:

- A small SRAM memory ($2^n \times 1$, with a value of n typically from 4 to 6) working as a lookup table (LUT), which allows any combinational function of its n inputs to be implemented. A LUT can be thought of as a way of storing the truth table of the combinational function in such a way that, when using the inputs of that function as address bits of the LUT, the memory bit storing the value of the function for each particular input combination can be read at the output of the LUT.

- A flip-flop whose data input is connected to the output of the LUT.
- A multiplexer (MUX) that selects as output of the LB either the flip-flop output or the LUT output (i.e., the flip-flop input). In this way, depending on the configuration of the MUX, the LB can implement either combinational or sequential functions.
- The inputs of the LB (i.e., of the LUT) and its output (i.e., of the MUX), connected to nearby IL.

In practice, actual LBs consist of a combination of several (usually two) of these basic LUT/flip-flop/MUX blocks (which are sometimes referred to as slices). They also often include specific logic to generate and propagate carry signals (both inside the LB itself and between neighbor LBs, using local carry-in and carry-out connections), resulting in the structure shown in Figure 1.6. Typically, in addition, the LUTs inside an LB can be combined to form a larger one, allowing combinational functions with a higher number of inputs to be implemented, thus providing designers with extra flexibility.

In addition, current FPGAs also include different kinds of specialized resources (described in detail in Chapters 2 through 5), such as memories and memory controllers, DSP blocks (e.g., MAC units), and embedded processors and commonly used peripherals (e.g., serial communication interfaces), among others. They are just mentioned here in order for readers to understand the ever-increasing application scope of FPGAs in a large variety of industrial control systems, some of which are highlighted in Section 1.5 to conclude this chapter.

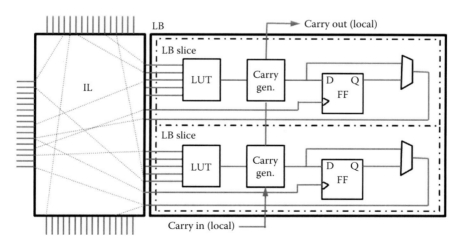

FIGURE 1.6
Example of two-slice LB and its connection to IL.

1.5 Applications and Uses of FPGAs

The evolution from "traditional" FPGA architectures, mainly consisting of basic standard reconfigurable building blocks (LBs and IL), to more feature-rich, heterogeneous devices is widening the fields of applicability of FPGAs, taking advantage of their current ability to implement entire complex systems in a single chip. FPGAs are not used anymore just for glue logic or emulation purposes, but have also fairly gained their own position as suitable platforms to deal with increasingly complex control tasks and are also getting, at a very fast pace, into the world of HPC.

This technological trend has also extended the applicability of FPGAs in their original application domains. For instance, emulation techniques are evolving into mixed solutions, where the behavior of (parts of) a system can be evaluated by combining simulation models with hardware emulation, in what is nowadays referred to as hardware-in-the-loop (HIL). Tools exist, including some of general use in engineering, such as MATLAB®, which allow this combined simulation/emulation approach to be used to accelerate system validation.

FPGAs are also increasingly penetrating the area of embedded control systems, because in many cases, they are the most suitable solution to deal with the growing complexity problems to be addressed in that area. Some important fields of application (not only in terms of technological challenges but also in terms of digital systems' market share) are in automated manufacturing, robotics, control of power converters, motion and machinery control, and embedded units in automotive (and all transportation areas in general)—it is worth noting that a modern car has some 70–100 embedded control units onboard. As the complexity of the systems to be controlled grows, microcontroller and DSPs are becoming less and less suitable, and FPGAs are taking the floor.

A clear proof of the excellent capabilities of current FPGAs is their recent penetration in the area of HPC, where a few years ago, no one would have thought they could compete with software approaches implemented in large processor clusters. However, computing-intensive areas such as big data applications, astronomical computations, weather forecast, financial risk management, complex 3D imaging (e.g., in architecture, movies, virtual reality, or video games), traffic prediction, earthquake detection, and automated manufacturing may currently benefit from the acceleration and energy-efficient characteristics of FPGAs.

One may argue these are not typical applications of industrial embedded systems. There is, however, an increasing need for embedded high-performance systems, for example, systems that must combine intensive computation capabilities with the requirements of embedded devices, such as portability, small size, and low-energy consumption. Examples of such applications are complex wearable systems in the range of augmented or virtual reality, automated driving vehicles, and complex vision systems for robots or in industrial plants. The Internet of Things is one of the main forces behind the trend to integrate increasing computing power into smaller and energy-efficient devices, and FPGAs can play an important role in this scenario.

Given the complexity of current devices, FPGA designers have to deal with many different issues related to hardware (digital and analog circuits), software (OSs and programming for single- and multicore platforms), tools and languages (such as HDLs, C, C++, SystemC, as well as some with explicit parallelism, such as CUDA or OpenCL), specific design techniques, and knowledge in very diverse areas such as control theory, communications, and signal processing. All these together seem to point to the need for super-engineers (or even super-engineering teams), but do not panic. While it is not possible to address all these issues in detail in a single book, this one intends at least to point industrial electronics professionals who are not specialists in FPGAs to the specific issues related to their working area so that they can first identify them and then tailor and optimize the learning effort to fulfill their actual needs.

References

Jones, D.H., Powell, A., Bouganis, Ch.-S., and Cheung, P.Y.K. 2010. GPU versus FPGA for high productivity computing. In *Proceedings of the 20th International Conference on Field Programmable Logic and Applications*, August 31 to September 2, Milano, Italy.

Kaeli, D. and Akodes, D. 2011. The convergence of HPC and embedded systems in our heterogeneous computing future. In *Proceedings of the IEEE 29th International Conference on Computer Design (ICCD)*, October 9–12, Amherst, MA.

Laprie, J.C. 1985. Dependable computing and fault tolerance: Concepts and terminology. In *Proceedings of the 15th Annual International Symposium on Fault-Tolerant Computing (FTCS-15)*, June 19–21, Ann Arbor, MI.

Shuai, C., Jie, L., Sheaffer, J.W., Skadron, K., and Lach, J. 2008. Accelerating compute-intensive applications with GPUs and FPGAs. In *Proceedings of the Symposium on Application Specific Processors (SASP 2008)*, June 8–9, Anaheim, CA.

2

Main Architectures and Hardware Resources of FPGAs

2.1 Introduction

Since their advent, microprocessors were for many years the only efficient way to provide electronic systems with programmable (user-defined) functionality. As discussed in Chapter 1, although their hardware structure is fixed, they are capable of executing different sequences of basic operations (instructions). The programming process mainly consists of choosing the right instructions and sequences for the target application.

Another way of achieving user-defined functionality is to use devices whose internal hardware resources and interconnects are not totally configured by default. In this case, the process to define functionality (configuration, as also introduced in Chapter 1) consists of choosing, configuring, and interconnecting the resources to be used. This second approach gave rise to the FPGA concept (depicted in Figure 2.1), based on the idea of using arrays of custom logic blocks surrounded by a perimeter of I/O blocks (IOBs), all of which could be assembled arbitrarily (Xilinx 2004).

From the brief presentation made in Section 1.4, this chapter describes the main characteristics, structure, and hardware resources of modern FPGAs. It is worth noting that it is not intended to provide a comprehensive list of resources, for which readers can refer to specialized literature (Rodriguez-Andina et al. 2007, 2015) or vendor-provided information. Embedded soft and hard processors are separately analyzed in Chapter 3 because of their special significance and the design paradigm shift they caused (as they transformed FPGAs from hardware accelerators to FPSoC platforms, contributing to an ever-growing applicability of FPGAs in many domains). DSP and analog blocks are other important hardware resources that are separately analyzed in Chapters 4 and 5, respectively, because of their usefulness in many industrial electronics applications.

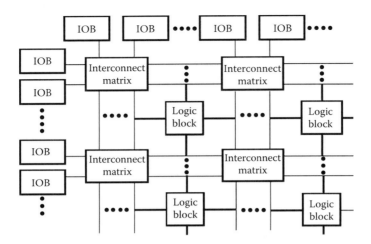

FIGURE 2.1
FPGA concept.

2.2 Main FPGA Architectures

The basic architecture of most FPGAs is the one shown in Figure 2.1, based on a matrix of configurable hardware basic building blocks (LBs,* introduced in Chapter 1) surrounded by IOBs that give FPGA access to/from external devices. The set of all LBs in a given device is usually referred to as "distributed logic" or "logic fabric." An LB can be connected to other LBs or to IOBs by means of configurable interconnection lines and switching matrices (IL, as also introduced in Chapter 1) (Kuon et al. 2007; Rodriguez-Andina et al. 2007, 2015).

In addition to distributed logic, aimed at supporting the development of custom functions, FPGAs include specialized hardware blocks aimed at the efficient implementation of functions required in many practical applications. Examples of these specific resources are memory blocks, clock management blocks, arithmetic circuits, serializers/deserializers (SerDes), transceivers, and even microcontrollers. In some current devices, analog functionality (e.g., ADCs) is also available. The combination of distributed logic and specialized hardware results in structures like the ones shown in Figure 2.2 (Xilinx 2010; Microsemi 2014; Achronix 2015; Altera 2015a).

---

* LBs receive different names from different FPGA vendors or different families from the same vendor (e.g., Xilinx, configurable logic block [CLB]; Altera, adaptive logic module [ALM]; Microsemi, logic element [LE]; Achronix, logic cluster [LC]), but the basic concepts are the same. This also happens in the case of IOBs.

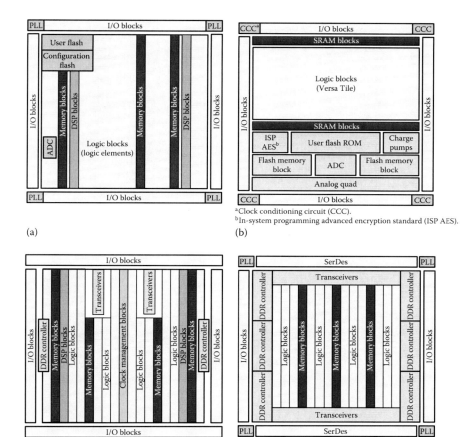

FIGURE 2.2
(a) Altera MAX 10, (b) Microsemi's Fusion, (c) Xilinx's Spartan-6, and (d) Achronix's Speedster22i HD architectures. *Note:* In December 2015, Intel Corporation acquired Altera Corporation. Altera now operates as a new Intel business unit called Programmable Solutions Group.

The main drawback of the matrix architecture, related to the number of IOBs required in a given device, affects complex FPGAs. As more distributed and specialized logic is included in a given device, more IOBs are also required, increasing cost. Another limitation of this architecture comes from the fact that power supply and ground pins are located in the periphery of the devices, and then voltage drops inevitably happen as supply/ground currents flow to/from the core from/to these pins. A third limitation is that the ability to scale specialized hardware blocks depends on the amount of distributed logic available in their vicinity.

To mitigate these limitations, vendors have developed column-based FPGA architectures (Xilinx 2006; Altera 2015b) like the one depicted in Figure 2.3.

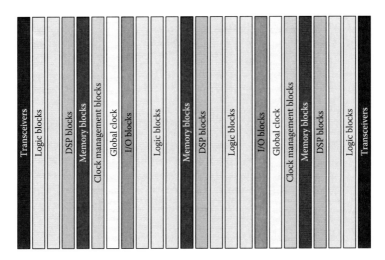

FIGURE 2.3
Column-based architecture.

First, in these architectures, there is no dependency of the number of IOBs on the amount of distributed and specialized logic because different types of resources are placed in dedicated, independent columns. This means that IOBs are located in their corresponding columns, and not just in the periphery, and the number of IOBs only depends on the number of I/O pins the vendor decides the device to have. This actually applies to any resource: if more resources of a given type are to be included in a device, the number of columns of such type is just increased. Power supply and ground pins are distributed throughout the whole chip area, thus minimizing signal integrity problems.

Column architectures are application oriented, because FPGAs with very different resources can be readily developed using chips with the same area and pin count, allowing the cost–performance trade-off to be optimized for each particular application.

A column architecture specifically targeting high-frequency/bandwidth applications is shown in Figure 2.4. In it, flip-flops ("Hyper-Registers") are placed in all interconnection segments and in all inputs of dedicated functional blocks, in addition to the usual locations in LBs and IOBs (as described in Section 2.3).

The availability of these flip-flops throughout the entire device allows design techniques such as retiming and pipelining to be more efficiently implemented (Hutton 2015). The use of such techniques reduces signal delay times, in turn allowing higher operation frequencies to be achieved. In more "classical" architectures, the implementation of these techniques must be done using the flip-flops of the distributed logic, which usually implies that

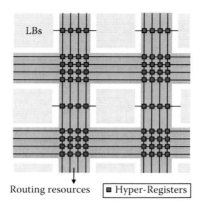

Routing resources   ■ Hyper-Registers

FIGURE 2.4
Altera's HyperFlex architecture.

an advantage cannot be taken from most of the resources of the LBs used and that delays are also higher, resulting in less efficient solutions.

2.3 Basic Hardware Resources

According to Figure 2.1, the basic hardware resources of FPGAs are LBs, IOBs, and interconnection resources.

2.3.1 Logic Blocks

LBs are intended to implement custom combinational and sequential functions. As a consequence, they mainly consist of reconfigurable combinational functions and memory elements (flip-flops/latches). The combinational part can be implemented in different ways (e.g., with logic gates or MUXs), but nowadays, lookup tables (LUTs, introduced in Chapter 1) are the most frequently used combinational elements of LBs. The differences among LBs from different vendors (or different FPGA families from the same vendor) basically refer to the number of inputs of the LUTs (which define the maximum number of logic variables the combinational function implemented in the LUT can depend on), the number of memory elements, and the configuration capabilities of the LB. Two sample LBs are shown in Figure 2.5, from Microsemi's IGLOO2 (Microsemi 2015a) and Achronix's Speedster22i HD1000 devices (Achronix 2015), respectively.

The complexity of LBs depends on the kind of applications a given FPGA family targets. The LB in Figure 2.5a corresponds to one of the simplest

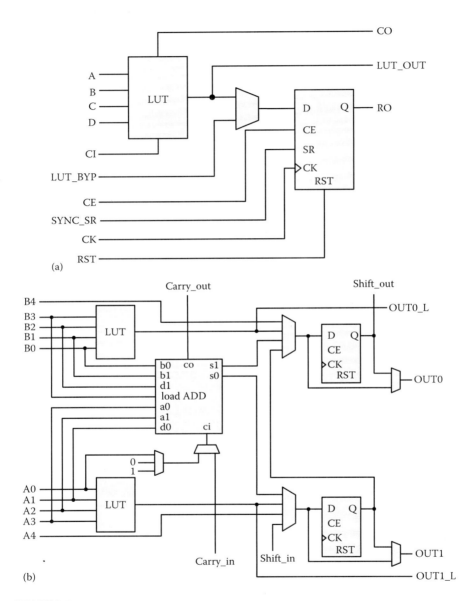

FIGURE 2.5
(a) LE from Microsemi's IGLOO2 and (b) heavy logic cluster from Achronix's Speedster22i HD1000 devices.

existing structures. It allows logic functions with up to four inputs to be implemented and includes just one flip-flop, which can be used either independently or to memorize the output of the LUT. In addition, specific lines (CIN and CO) allow carry signals to be propagated from the LB to a contiguous one. Carry propagation chains are a typical resource in any LB, which simplifies the efficient implementation of widely used functions such as counters or adders.

On the other hand, the LB in Figure 2.5b is more complex. It consists of two four-input LUTs, one embedded adder, and two flip-flops. Using the corresponding MUX, each LUT can implement a single five-input function. In addition, LUTs and MUXs can be combined to implement certain six- to nine-input functions (Achronix 2015). Embedded adders, supporting 2 bit operands, allow addition-based computations to be accelerated. Finally, the availability of two flip-flops per LB targets register-intensive solutions, such as pipelining. The remaining elements, mainly MUXs, provide configurability, enabling many different combinations of the other resources to be configured as well as easing routability of input and output signals.

In the vast majority of FPGAs, basic LBs are grouped in blocks of higher hierarchy sharing specific interconnection resources, allowing more complex functions to be implemented with short additional delays. Also, flip-flops can be combined to create shift registers, delay lines, or distributed* memories (ROM, single- or dual-port RAM, or FIFOs).

A specific solution (Achronix picoPIPE) aimed at the implementation of pipelined datapaths without the need for adding intermediate flip-flops/registers (therefore avoiding modifications to be required in the original, nonpipelined logic structure) is shown in Figure 2.6 (Achronix 2015). It is

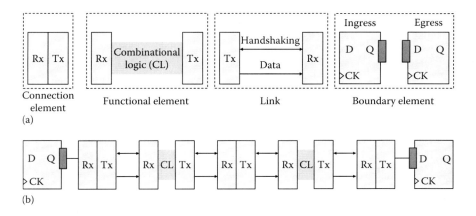

FIGURE 2.6
(a) Achronix picoPIPE building blocks and (b) pipeline stages.

---

* So called because they are built using distributed logic.

based on a handshake-controlled, asynchronous propagation of data, instead of the usual clock-synchronized propagation of conventional FPGA logic. There are four basic building blocks in Figure 2.6:

- Functional elements, which not only implement the target combinational logic but also handshake data input and output, thus ensuring only valid data are propagated.
- Connection elements, which provide resources for both connectivity and storage (flip-flops). Therefore, they can act as simple data repeaters or as registers, enabling either asynchronous or synchronous computations to be implemented.
- Links to communicate functional elements.
- Boundary elements, used as interface between picoPIPE and conventional FPGA logic. Data entering (exiting) the picoPIPE fabric must pass through ingress (egress) boundary elements.

The use of pipeline stages like the ones shown in Figure 2.6 allows any logic function to be implemented using the same logic structure it would have in a nonpipelined conventional FPGA implementation and implicitly add pipeline stages as required to shorten propagation delay times, reaching operating frequencies up to 1.5 GHz (Achronix 2008). A sample comparison between conventional and picoPIPE implementations is depicted in Figure 2.7 (interconnection resources are described in Section 2.3.3).

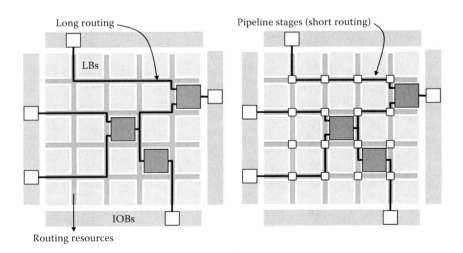

FIGURE 2.7
Comparison between conventional and picoPIPE implementations.

2.3.2 I/O Blocks

IOBs serve as links between device pins and internal resources. Their main elements are programmable bidirectional buffers and flip-flops to synchronize I/O and configuration signals, as shown in Figure 2.8 (Altera 2012; Xilinx 2014a; Microsemi 2015b).

Similar to the case of LBs, IOBs with different levels of complexity are available in the different families of current FPGA devices. However, they all share some common features:

- Input data can either be directly connected to the internal resources or pass through a memory element. Similarly, output data can pass through a memory element or bypass it.
- Memory elements can be configured as flip-flops or latches.
- Bidirectional buffers support different voltage levels (1.2, 1.5, 1.8, 2.5, 3.0, and 3.3 V) and different I/O standards (single-ended, differential, or voltage-referenced). The most commonly available ones are low-voltage TTL, low-voltage CMOS, stub series-terminated logic (SSTL), differential SSTL, high-speed transceiver logic (HSTL), differential HSTL, high-speed unterminated logic (HSUL), and low-voltage differential signaling (LVDS).

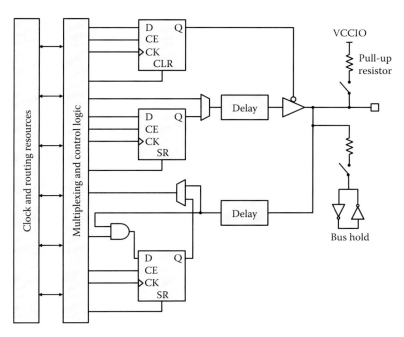

FIGURE 2.8
Bidirectional IOB.

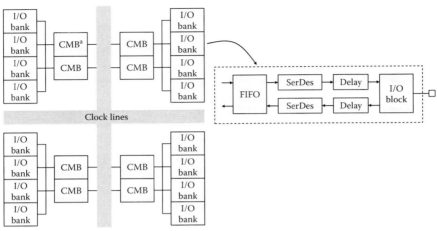

ᵃClock management block (CMB).

FIGURE 2.9
I/O banks.

IOBs are grouped in banks sharing common resources and, usually, configuration details (Altera 2015b; Microsemi 2015b; Xilinx 2015a), as shown in the example in Figure 2.9. Each bank can be configured to support a different I/O standard (in some advanced device families, several standards can be combined in the same bank). Since each standard has its own specifications for voltages, currents, types of buffer, and types of termination, the ability to adapt the same FPGA to simultaneously use several I/O standards allows it to be connected to circuits operating under different electrical conditions (e.g., different power supply voltages) without the need for external conditioning circuitry. This simplifies PCB design and decreases design time, in turn significantly reducing cost.

- Programmable control of the output current for some I/O standards. This feature allows the output buffer of the IOB to comply with the I_{OH} and I_{OL} specifications of the configured standard, reducing simultaneous switching output effects and, in turn, noise.

- Programmable control of the output slew rate (rising and falling), which can be independently configured for each pin at different levels (available levels vary among devices), for example, *slow* or *fast*. For outputs operating at high frequencies, fast configurations should be used, but attention must be paid to possible signal reflection problems and noise transients during switching.

- Programmable pull-up and pull-down resistors.

- Programmable delay lines to control setup and hold times in input flip-flops/latches and clock-to-output propagation times in output flip-flops/latches or to delay input clock signals.
- Support for double data rate (DDR) I/O. This implies IOBs include at least two input and two output flip-flops and two clock signals with a 180° phase shift between them. Flip-flops can be configured to capture data in the same edge of both clocks or in opposite edges, thus allowing different data alignment modes to be implemented.
- Programmable output differential voltage (V_{OD}). This allows the right trade-off between voltage margin of the external circuit (which increases for higher V_{OD}) and FPGA power consumption (which decreases for lower V_{OD}) to be achieved for each particular application.

As may be noticed in Figure 2.9, IOBs can include specialized elements in addition to the ones mentioned earlier. These functionalities may only be available in the most complex (and expensive) devices. Two of the most useful ones, SerDes blocks and FIFO memories, are described in Sections 2.3.2.1 and 2.3.2.2.

2.3.2.1 SerDes Blocks

SerDes blocks are serial–parallel (input deserializer) and parallel–serial (output serializer) conversion circuits to interface digital systems with serial communication links. They significantly ease the implementation of systems with high data transfer rate requirements, such as in video applications, high-speed communications, high-speed data acquisition, and serial memory access.

In some FPGAs, SerDes blocks can only work with differential signals; that is, they can only be used when the corresponding IOBs are configured to work in a differential I/O standard. In other devices, they can work with both single-ended and differential signals.

SerDes can support different operating modes and work at different data transfer rates (e.g., single data rate or DDR modes). In some cases, they can be connected in chain to achieve higher rates.

A SerDes block from Altera's Arria 10 family (Altera 2015b) is shown in Figure 2.10. The upper part corresponds to the output serializer, whereas the input deserializer is depicted in the lower part. One of the most critical issues in the design of this kind of circuits is related to the requirements imposed on clock signals. Due to this, some FPGAs include dedicated clock circuits (independent from global clock signals) in their SerDes blocks (e.g., I/O phase-locked loop [PLL] in Figure 2.10).

The input deserializer usually includes a *bit slip* module to reorder the sequence of input data bits. This feature can be used, for instance, to correct

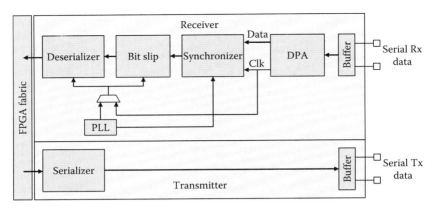

FIGURE 2.10
Altera's Arria 10 family SerDes block.

skew effects among different input channels (like in Altera's Arria 10 devices) or to detect the training patterns used in many communication standards (like in Xilinx' Series 7 devices).

In some FPGA families (e.g., Altera's Arria 10), the input deserializer also includes a dynamic phase alignment circuit (DPA in Figure 2.10) that allows input bits to be captured with minimum skew with regard to the deserializer's clock signal. This is accomplished by choosing as clock signal, among several of them with different phases, the one with minimum phase shift with regard to input bits.

2.3.2.2 FIFO Memories

I/O FIFO memories are available in some of the most advanced FPGAs (like Xilinx' Series 7 devices). They are mainly intended to ease access to external circuits (such as memories) in combination with SerDes or DDR I/O resources, but can also be used as *fabric* (general-purpose) FIFO resources.

2.3.3 Interconnection Resources

Interconnection resources form a mesh of lines located all over the device, aimed at connecting the inputs and outputs of all functional elements of the FPGA (LBs, IOBs, and specialized hardware blocks—described in Section 2.4). They are distributed in between the rows and columns of functional elements, as shown in Figure 2.11.

Interconnect lines are divided into segments, trying to achieve the minimum interconnect propagation delay times according to the location of the elements to be connected. There are segments with different lengths, depending on whether they are intended to connect elements located close to each other (short segments) or in different (distant) areas of the device (long segments).

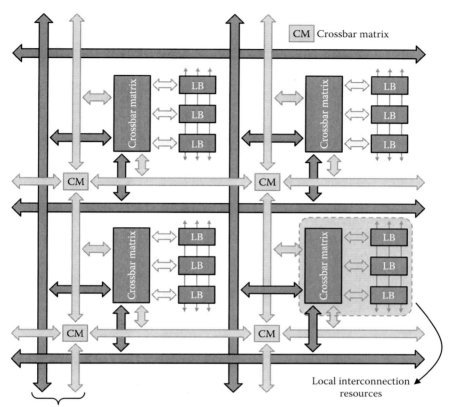

General FPGA interconnection resources:
short segments, long segments, and global lines

FIGURE 2.11
General and local FPGA interconnection resources.

For specific signals expected to have high fan-out, for example, clock, set, or reset (*global*) signals, special lines are used, covering the entire device or large regions in it. For instance, clock signals have a dedicated inter-connection infrastructure, consisting of *global* lines and groups of *regional* lines, each group associated with a specific area of the device, as discussed in Section 2.4.1. The stacked silicon interconnect technology used in some Xilinx' Virtex-7 devices allows performance to be improved, thanks to ultra-fast clock lines and a fast type of interconnection resource called superlong lines (Saban 2012).

In order for a particular interconnection to be built, several segments are connected in series by means of crossbar matrices.* Since LBs are the most

---

* Like in the case of LBs and IOBs, the terminology for interconnection resources varies depending on the FPGA vendor.

abundant resources and have a significant number of input and output signals (as can be noticed in Figure 2.5), they are usually first connected to a dedicated crossbar matrix shared by several of them (*local* interconnection resources) and, from it, to the *general* FPGA interconnection resources (Xilinx 2014b; Achronix 2015; Altera 2015a; Microsemi 2015c), as shown in Figure 2.11.

Interconnection delays are a critical factor in the performance of FPGA designs. They depend on the type of resources used, the number of matrices to be crossed, and the distance to be traveled by signals. Because of this, the assignment (*placement*) of the functional blocks of a given circuit to the best possible actual hardware resources in the FPGA is a key factor in achieving the best possible performance. Software design tools should provide suitable placements, but in some cases (in particular for complex designs requiring the use of most of the available hardware resources), the best performance can only be obtained with some designer intervention at the device floorplan level (or, if feasible, by using higher-end, more expensive, devices).

2.4 Specialized Hardware Blocks

Several types of specialized hardware blocks are available in most current FPGAs, but not all of them are available in all devices and their number varies from one device to another. Actually, the type and number of specialized hardware resources included in a given device determine its target application domain. Some of the most usual specialized hardware resources—clock management blocks, memory blocks, and transceivers—are described in the following sections. As stated in Section 2.1, because of their special significance, embedded soft and hard processors, as well as DSP and analog blocks, are separately analyzed in Chapters 3 through 5, respectively.

2.4.1 Clock Management Blocks

The generation, control, and quality of clock signals are among the most important problems to be faced in the design of complex digital systems, particularly in the case of multirate systems or those requiring very fast data transfer rates, where synchronization among the different parts of the system is a critical issue.

Regarding clock management, FPGAs are divided into regions designed to minimize clock propagation delays within them. A set of dedicated clock input pins is assigned to each region, together with resources to manage and distribute clock signals (Actel 2010; Achronix 2015; Altera 2015c; Microsemi 2015d; Xilinx 2015b), as shown in Figure 2.12. The number of

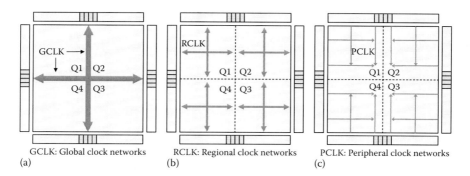

FIGURE 2.12
(a) Global, (b) regional, and (c) peripheral clock networks in Altera's Stratix V devices.

clock regions varies depending on the device size. Global clock lines also exist, as well as other clock lines that connect adjacent clock regions. In some FPGAs, it is possible to execute a *clock power down*, "disconnecting" global or regional clock signals to reduce power consumption. When a clock line is powered down, all logic associated with it is also switched off, further reducing power consumption.

In order to reduce the problems associated with clock signals as well as the number of external oscillators needed, FPGAs include clock management blocks (CMBs),* based on either PLLs or delay-locked loops (DLLs). These CMBs are mainly used for frequency synthesis (frequency multiplication or division), skew reduction, and duty cycle/phase control of clock signals.

Each CMB is associated with one or several dedicated clock inputs, and in most devices, it can also take as input an internal global clock signal or the output of another CMB (chain connection). Chain connections allow dynamic range to be increased for frequency synthesis (both for frequency multiplication and division).

For optimized performance, CMBs are physically placed close to IOBs and are connected to them with dedicated resources. Therefore, in matrix architectures, such as Microsemi's IGLOO2 (Figure 2.13a), CMBs are placed in (or close to) the periphery, where IOBs are located. In column-based architectures, such as Xilinx' Series 7 (Figure 2.13b), CMBs are placed in specific columns, regularly distributed all over the device, but always next to IOBs columns. It should be noted that, in most FPGAs, CMBs are also used to generate the clock signals used by SerDes blocks and transceivers.

Although the functionality of CMBs is similar regardless of the vendor/family of devices, their hardware structures are quite diverse. As mentioned

---

* Again, different vendors use different names. Xilinx, clock management tiles or digital clock managers; Altera, PLLs or fractional PLLs; Microsemi, clock conditioning circuitry; Achronix, global clock generator.

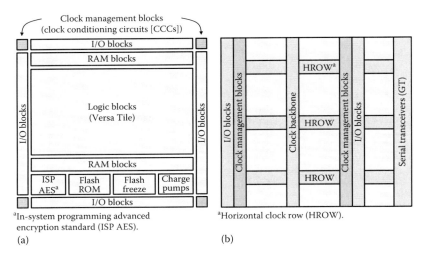

FIGURE 2.13
Location of CMBs in (a) matrix and (b) column-based architectures.

before, some of them are based on DLLs (digital solution), but most current devices use PLLs (analog solution). Basic PLLs work with integer factors for frequency synthesis (integer PLLs), but in the most advanced devices, fractional PLLs (capable of working with noninteger factors) are also available.

The structure of an integer PLL is depicted in Figure 2.14. Its main purpose is to achieve perfect synchronization (frequency and phase matching) between its output signal and a reference input signal (Barrett 1999). Its operation is as follows: The phase detector generates a voltage proportional to the phase difference between the feedback and reference signals. The low-pass filter averages the output of the phase detector and applies the resulting signal to a voltage-controlled oscillator, whose resonant frequency (the output frequency) varies accordingly. In this way, the output frequency is dynamically adjusted until the phase detector indicates that the feedback

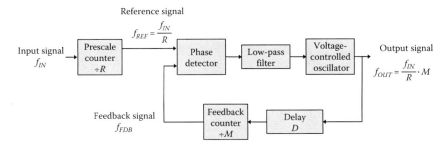

FIGURE 2.14
Block diagram of an integer PLL.

and reference signals are in phase. At this point, the PLL is said to have reached the phase-lock condition.

In case the feedback loop is a direct connection of the output signal to the input of the phase detector (i.e., there is neither a delay block nor a feedback counter), in steady state, the frequency and phase of the output signal follow those of the reference input signal.

If the delay block is included in the feedback loop, in order for the phase-lock condition to be achieved, the phase of the output signal must lead that of the reference signal by an amount equal to the delay in the feedback loop (Gentile 2008).

Similarly, if the counter is included in the feedback loop, the frequency of the output signal will be M times the frequency of the reference signal. In this way, the PLL acts as a frequency multiplier by an integer factor M. By simply varying M, the output frequency can be adjusted to any integer multiple of the reference frequency (within the operating limits of the circuit). If the reference frequency is obtained by dividing the frequency of an input signal by an integer scaling factor R (prescale counter in Figure 2.14), the output frequency will also be divided by R; that is, the effective multiplying factor would be M/R.

Therefore, the relatively simple structure in Figure 2.14 allows CMBs to synthesize multiple frequencies from an input clock signal, control the phase of the output signal, and eliminate skew by synchronizing the output signal with the input reference signal. As an example of an actual circuit (Actel 2010), the one in Figure 2.15 provides five programmable dividing counters (C_1–C_5) that can generate up to three signals with different frequencies. There are two delay lines in the feedback loop (one fixed and one programmable) that can be used to advance the output clock relative to the input clock. Another five lines are available to delay output signals.

In spite of their simplicity and usefulness, PLLs based on integer divisions have two main drawbacks:

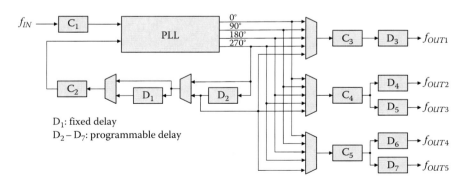

FIGURE 2.15
Structure of Microsemi's integer PLL.

- When multiplying frequency by M, phase noise (jitter in time domain) in the output signal increases $20 \cdot \log(M)$. This effect may be mitigated by using a higher reference frequency (which would imply the use of a lower value of M to obtain the same output frequency), but this is not always possible because the reference frequency defines the frequency resolution of the PLL, and for some applications, it is a design specification (Barrett 1999; Texas Instruments 2008).

- The cutoff frequency of the low-pass filter must be lower enough than the reference frequency. For lower cutoff frequencies, the acquisition (or lock) time of the PLL increases. This is the time needed for the PLL to reach steady state (i.e., to synchronize) after power on, reset, or the reconfiguration of its operating parameters (Barrett 1999).

Fractional PLLs have a better behavior than integer ones in terms of phase noise and acquisition time. Their (*fractional*) frequency resolution is a fraction F of the reference frequency. This means that input frequency can be F times the frequency resolution, resulting in lower phase noise and acquisition time.

Fractional PLLs are based on the use of a divider by $M + K/F$ in the feedback loop, where K is the *fractional multiply factor*. As explained earlier, this would be the frequency multiplying factor of the PLL unless a prescale frequency divider by R is applied to the input signal.

There are two hardware approaches to obtain a fractional PLL, which are used in different FPGA devices. The simplest one uses an accumulator to dynamically modify the frequency division in the feedback loop, in such a way that in K out of F cycles of the reference signal, the dividing factor is $M + 1$, and in $F - K$ cycles, the frequency is divided by M, resulting in an average dividing factor equal to $[(M + 1)K + M \cdot (F - K)]/F = M + K/F$.

This solution adds spurious signals (instantaneous phase errors in the time domain) to the output frequency. Although they can be mitigated by using analog methods, a better solution is achieved by using a different hardware structure for the fractional PLL. In this second approach, based on a delta-sigma modulator, digital techniques are used to more efficiently reduce phase noise and spurious signals (Barrett 1999; Texas Instruments 2008).

As an example, the CMB shown in Figure 2.16 (which combines integer and fractional PLLs) uses a delta-sigma modulator associated with the feedback frequency divider (Altera 2015b). It also includes several output dividers (C_0–C_n) to generate output clock signals of different frequencies, as well as an input clock switch circuit to select the reference signal. Reference signals may be the same (clock redundancy) or have different frequency (for dual-clock-domain applications). The input and output

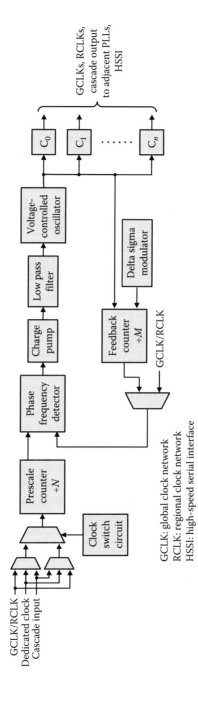

FIGURE 2.16
Integer/fractional PLL from Altera Arria 10 family.

GCLK: global clock network
RCLK: regional clock network
HSSI: high-speed serial interface

signals of this CMB can be connected to global or regional clock lines, to external clock pins, or to other CMBs.

In the CMBs of any current FPGA, the feedback signal can be obtained from different sources and be routed through different paths. The way of doing it depends on the target functionality, as described in the following (it must be noted that not all devices provide all these possibilities):

- Minimize the length of the feedback path to reduce output jitter.*
- Compensate skew in the clock network used to generate the output of the CMB, which can be generated using either internal or external feedback. In the first case, feedback comes from a global or regional clock line, compensating internal device delays; whereas in the second case, feedback comes from a device pin, compensating delays at the board level.
- Generate zero-delay buffer clocks (Gentile 2008). When the signal generated by the CMB is connected to an external clock pin, it may be important to compensate the propagation delays introduced by this pin and the external connections, in order to ensure that the clock signal reaching the external device is synchronized with the CMB's reference signal.
- Ensure the phase in the data and clock inputs of the memory elements in IOBs is the same as the phase of the same signals when they reach the device pins; that is, the pin-to-register-input delays of these signals are the same.
- Ensure this equality of delays from input pins also for the clock and data input signals of SerDes blocks.

In spite of their similar functionalities, there are many differences among CMBs from different FPGA families in terms of input and output frequency ranges, frequency/phase synchronization ranges, access to interconnection resources, types of signals they can generate (e.g., single-ended, differential), the number of outputs, possible values of frequency multiplying and dividing factors, fixed/variable/programmable delay, and so on. There are obviously also differences in the control signals, but at least two of them are present in all devices: reset, to initialize the CMB, and locked, whose activation validates the output signal (i.e., indicates frequency and/or phase synchronization has been achieved). The combination of both signals allows the correct behavior of the CMB to be checked and recovered if needed. If synchronism is lost, the locked signal will be deactivated. As a response, a reset can be launched for synchronism to be recovered. This process can be automatically executed in some FPGAs.

---

* Some vendors refer to this as jitter filter.

Although from all the previously mentioned issues, it may seem that it is difficult for the user to deal with the many different configuration parameters and operating modes of CMBs, actually this is not the case. Software design tools usually offer IP blocks whose user interfaces require just a few values to be entered and then configuration parameters are automatically computed.

2.4.2 Memory Blocks

Most digital systems require resources to store significant amounts of data. Memories are the main elements in charge of this task. Since memory access times are usually much longer than propagation delays in logic circuits, memories (in particular external ones) are the bottleneck of many systems in terms of performance. Because of this, FPGA vendors have always paid special attention to optimizing logic resources so that they can support, in the most efficient possible way, the implementation of internal memories.

Since combinational logic, LUTs, and flip-flops are available in LBs, internal memories can be built by combining the resources of several LBs, resulting in the so-called distributed memory. However, in order for distributed memory to be more efficient, LBs may be provided with resources additional to those intended to support the implementation of general-purpose logic functions, such as additional data inputs and outputs, enable signals, and clock signals. Because this implies LBs to be more complex and, in addition, it makes no sense to use all LBs in an FPGA to build distributed memories, usually only around 25%–50% (depending on the family of devices) of the LBs in a device are provided with extra resources to facilitate the implementation of different types of memories: RAM, ROM, FIFO, shift registers, or delay lines (Xilinx 2014b; Altera 2015c). The structures of a "general-purpose" LB and another one suitable for distributed memory implementation can be compared in Figure 2.17.

As FPGA architectures evolved to support the implementation of more and more complex digital systems, memory needs increased. As a consequence, vendors decided to include in their devices dedicated memory blocks, which in addition use specific interconnection lines to optimize access time. They are particularly suitable for implementing "deep" memories (with a large number of positions), whereas distributed memory is more suitable for "wide" memories (with many bits per position) with few positions, shift registers, or delay lines.

In current FPGAs, both distributed memory and dedicated memory blocks support similar configurations and operating modes. Dedicated memory is structured in basic building blocks of fixed capacity, which can be combined to obtain deeper (series connection) or wider (parallel connection) memories. The possible combinations depend on the target type of memory and on the operating mode. The capacity of the blocks largely varies even

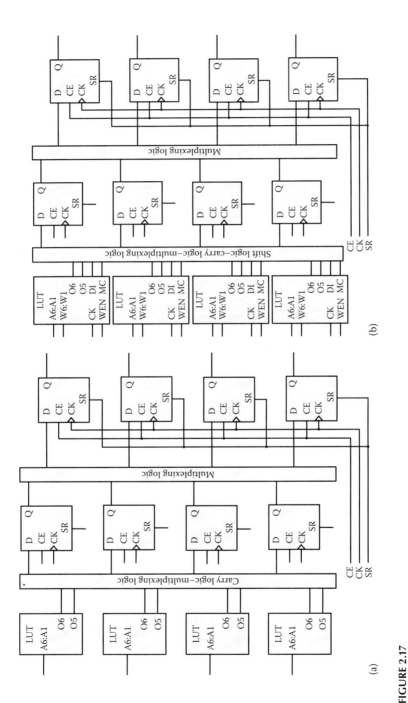

FIGURE 2.17
LBs from Xilinx' Series 7 devices: (a) general purpose and (b) oriented to distributed memory implementation.

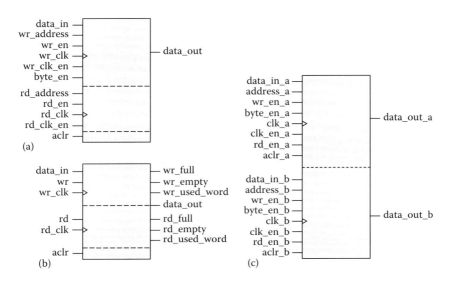

FIGURE 2.18
Sample Altera's Cyclone III memory modes: (a) simple dual-port block RAM, (b) FIFO, and
(c) true dual-port block RAM.

among devices of the same family (Altera 2012; Xilinx 2014c; Achronix 2015;
Microsemi 2015c). The most common configurations (some of which can be
seen in the sample case in Figure 2.18) are

- Single-port RAM, where only one single read or write operation can
  be performed at a time (each clock cycle)
- Simple dual-port RAM, where one read and one write operation can
  be performed simultaneously
- True dual-port RAM, where it is possible to perform two write operations, two read operations, or one read and one write operation
  simultaneously (and at different frequencies if required)
- ROM, where a read operation can be performed in each clock cycle
- Shift register
- FIFO, either synchronous (using one clock for both read and write
  operations) or asynchronous (using two independent clocks for read
  and write operations). They can generate status flags ("full," "empty,"
  "almost full," "almost empty"; the last two are configurable).

In dual-port memories, usually word width can be independently configured
for each port. In some cases, input and output word widths can also be
independently configured for the same port, which eases the efficient implementation of content-addressable memories. Configurations cannot be
arbitrary, but have to be chosen from a predefined set.

Several clock modes can be used in FPGA memories (some of which are mentioned earlier), but not all modes are supported in all configurations:

- *Single clock*: All memory resources are synchronized with the same clock signal.
- *Read/write*: Two different clocks are used for read and write operations, respectively.
- *Input/output*: Uses separate clocks for each input and output port.
- *Independent clocks*: Used in dual-port memories to synchronize each port with a different clock signal.

Some memory blocks support error detection or correction using parity bits or dedicated error correction blocks (Xilinx 2014c; Altera 2015b), as shown in Figure 2.19. These are complementary functionalities that can be configured from the software design tools.

Regarding parity, depending on data width, one or more parity bits may be added to the original binary combination. In some FPGAs, parity functions are not implemented in dedicated hardware, but have to be built from distributed logic. In Xilinx' Series 7 devices, parity is one of the possibilities offered by the error correction code (ECC) encoder. The circuit in Figure 2.19 cannot be used with distributed memory. It can exclusively be associated

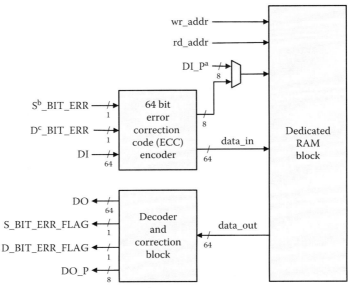

[a]Parity (P).   [b]Single-bit error (S).   [c]Double-bit error (D).

FIGURE 2.19
Error correction resources in Xilinx' Series 7.

with dedicated memory blocks, in particular with simple dual-port and FIFO configurations. It allows single-bit errors to be detected and corrected or double-bit errors to be detected. Output signals are available to flag the occurrence of an error and indicate whether or not it could be corrected.

Dedicated memory blocks based on SRAM cells can be found in all current FPGAs. In some devices, flash memories with read/write capabilities are also available (Microsemi 2014). Their main advantage comes from the fact of being nonvolatile, and their main drawback is that they require more control signals than SRAM-based ones, therefore making their control from the FPGA fabric more complex. ECC blocks are also available for this kind of memories.

The addition of memories to FPGA designs is facilitated by software design tools, which automatically partition the memory blocks defined by the designer and assign them to the memory blocks available in the target device, according to the operation modes specified and the design constraints regarding area and speed. Memory contents can also be initialized with the help of the design tools, which allow the contents of text files (where the values to be initially stored in the memories are described with a predefined syntax) to be included in the configuration bitstream.*

2.4.3 Hard Memory Controllers

In many FPGA applications, a huge amount of data has to be handled, but there is not enough embedded memory available for that. In such cases, external memory has to be used, and the corresponding memory controller needs to be implemented in the FPGA. Since there exist a wide variety of memories, the required interfaces are also very diverse, from simple parallel or serial interfaces (such as Serial Peripheral Interface [SPI], Inter-Integrated Circuit [I²C], and Universal Serial Bus [USB]) to much more complex ones (e.g., DDR).

To address this issue, FPGA vendors offer different soft[†] IP core-based solutions. However, these do not provide good-enough performance when dealing with very large memories (up to the GB range) or very fast operation requirements (hundreds of MHz or even GHz). This is the reason why FPGA vendors are including hard memory controllers in their most current devices. For instance, Arria V and 10 families from Altera include dedicated hardware for access control to external DDR/DDR2/DDR3/DDR4 SDRAM memories (Figure 2.20). Spartan-6 and Virtex-6 families from Xilinx also include DDR3 hard memory controllers, enhanced in Series 7 families of devices and extended in the UltraScale family to support DDR4 memories.

---

* FPGA configuration issues are analyzed in detail in Chapter 6.
[†] The functionality of soft cores is implemented using resources of the FPGA fabric.

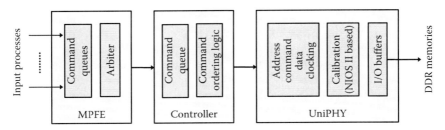

FIGURE 2.20
Arria 10 hard memory controller.

Two types of hard DDR/DDR2/DDR3 memory controllers are available in Microsemi SmartFusion2 devices, one of them accessible from the FPGA fabric and the other from an embedded ARM Cortex-M3 core* (so it cannot then be considered an FPGA resource, but rather one of the core). MachXO2, LatticeXP2, and LatticeECP2/M families from Lattice include circuitry allowing DDR/DDR2 memory interfaces to be implemented, whereas LatticeECP3, ECP5, and ECP5-5G families also support DDR3 memory interfaces.

Compared with soft IP core-based solutions, hard controllers achieve lower latencies and higher access frequencies. They support different data widths, reordering of commands and data for out-of-order execution, definition of priorities for reduced latency, streaming read or write operations for massive data transfer, burst modes, operation modes for continuous access to random sequences of memory addresses, multiport interfaces, low power consumption modes, user-controlled partial refresh cycles for reduced consumption, and error-correcting algorithms.

Let us consider the sample controller in Figure 2.20 (Altera 2016), consisting of three main building blocks (all of them physically located in the I/O banks of the devices):

- The physical layer interface (UniPHY) directly interacts with the I/O pins and is in charge of ensuring an adequate timing between the controller and the external memory. One of the main problems of external memory interfaces is the skew among data lines due to PCB routing. This problem is particularly significant for wide, high-speed buses. UniPHY mitigates this problem by means of configurable delay chains, which allow the delay associated with each I/O pin to be independently adjusted so as to align all data in the bus.

- The memory controller is in charge of maximizing bandwidth, through efficient control of the commands for external memory. It uses two main strategies for that, namely, reordering commands to take advantage of idle/dead cycles and reordering data and commands to

---

* As stated in Section 2.1, embedded soft and hard processors are separately analyzed in Chapter 3.

group read or write commands so that they are executed together, minimizing bus turnaround time.

- The multiport front end (MPFE) manages the access of multiple processes (read or write transactions) implemented in the FPGA fabric to the same hard external memory interface. In Arria 10 devices, it is a soft IP core.

2.4.4 Transceivers

A key factor for the success of FPGAs in the digital design market is their ability to connect to external devices, modules, and services in the same PCB, through backplane, or at long distance. In order to be able to support applications demanding high data transfer rates, the most recent FPGA families include full-duplex transceivers, compatible with the most advanced industrial serial communication protocols (Cortina Systems and Cisco Systems 2008; PCI-SIG 2014). Data transfer rates up to 56 Gbps can be achieved in some devices, and the number of transceivers per device can be in excess of 100 (e.g., up to 144 in Altera's Stratix 10 GX family and up to 128 in Xilinx's Virtex UltraScale + FPGAs). Some of the supported protocols are as follows:

- Gigabit Ethernet
- PCI express (PCIe)
- 10GBASE-R
- 10GBASE-KR
- Interlaken
- Open Base Station Architecture Initiative (OBSAI)
- Common Packet Radio Interface (CPRI)
- 10 Gb Attachment Unit Interface (XAUI)
- 10GH Small Form-factor Pluggable Plus (SFP+)
- Optical Transport Network OTU3
- DisplayPort

Transceivers are complex circuits, whose architectures vary among solutions from different FPGA vendors (as can be seen in Figure 2.21), in particular regarding generation and management of clock signals (Altera 2014, 2015d; Xilinx 2014d, 2015c; Achronix 2015; Jiao 2015; Microsemi 2015b). Anyway, they can be basically divided in two parts, namely, transmitter and receiver, each one in turn consisting of two main blocks (depicted in Figure 2.22 for the case of Altera's Stratix V devices): physical medium attachment (PMA) and physical coding sublayer (PCS).

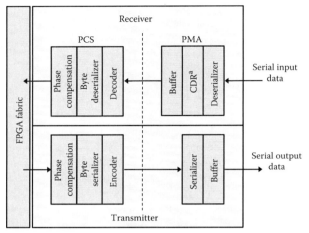

^aClock data recovery (CDR).

(a)

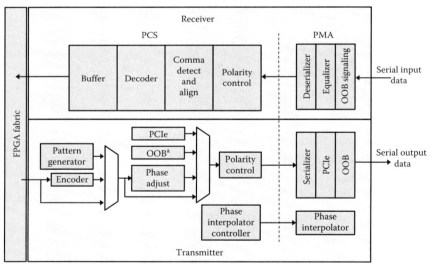

^aOut-of-band (OOB) sequences of the serial ATA (SATA).

(b)

FIGURE 2.21
Transceivers from (a) Xilinx' Series 7 and (b) Altera's Stratix V families.

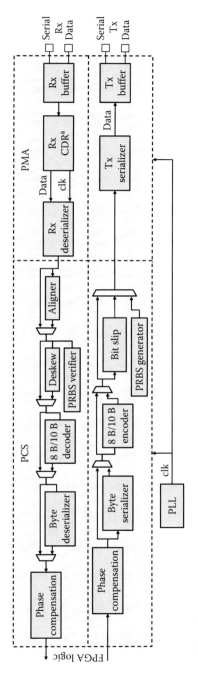

FIGURE 2.22
Altera's Stratix V transceiver: PMA (right) and PCS (left).

[a] Clock data recovery (CDR).

Data flows are as follows: In the receiver, serial input data enter the PMA block, whose output is applied to the PCS block, and finally information reaches the FPGA fabric. In the transmitter, output data follow a similar path, but in the opposite direction, from the FPGA fabric to the output of the PMA.

Given the high complexity of these blocks and taking into account that the detailed analysis of communication protocols is totally out of the scope of this book, only the main functional characteristics shared by most FPGA transceivers are described in the following text.

The receiver's PMA consists at least of an input buffer, a clock data recovery (CDR) unit, and a deserializer:

- The input buffer allows the voltage levels and the terminating resistors to be configured in order for the input differential terminals to be adapted to the requirements of the different protocols. It supports different equalization modes (such as continuous time linear equalization or decision feedback equalization) aimed at increasing the high-frequency gain of the input signal to compensate transmission channel losses.
- The CDR unit extracts (recovers) the clock signal and the data bits from incoming bitstreams.
- The deserializer samples the serial input data using the recovered clock signal and converts them into words, whose width (8, 10, 16, 20, 32, 40, 64, or 80 bits) depends on the protocol being used.

In the transmitter side, the PMA is in charge of serializing output data and sending them through a transmission buffer. This buffer includes circuits to improve signal integrity in the high-speed serial data being transmitted. Features include pre- and post-emphasis circuits to compensate losses, internal terminating circuits, or programmable output differential voltage, among others.

PCSs (both in the transmitter and in the receiver) can be considered as digital processing interfaces between the corresponding PMA and the FPGA fabric. Their main tasks are as follows:

- Encode (decode) data to be transmitted (being received) to support a variety of standard or proprietary coding solutions (8 B/10 B, 64 B/66 B, 64 B/67 B).
- Align serial input data to symbol boundaries (receiver).
- Generate (transmitter) or detect (receiver) the standard patterns (pseudo-random bit sequences [PRBS]) used to check signal integrity in high-speed serial links.

In addition, since transceivers use several clock domains, PCSs usually include deskew circuits (such as the ones described in Section 2.4.1) to align the phase of the different clock signals, as well as circuits to compensate small frequency variations between the external transmitter and the local receiver.

Depending on the operating mode or the used protocol, the PCS block may not be used. Actually, not all FPGA transceivers include this block. Some devices, in contrast, include transceivers with different types of PCS blocks, supporting different serial data transfer rates.

Finally, to ensure integrity of the transmitted data, transceivers must be calibrated before they start to operate. Transceivers in some devices (e.g., Altera's Stratix 10) include circuits that automatically perform the calibration process at power on.

Like in the cases of clock management and memory blocks, although transceiver configuration is in principle a complex task, software design tools provide resources to automatically obtain *wrappers* that allow transceivers to be configured from either predefined models of industrial standards or user-defined custom protocols.

2.4.4.1 PCIe Blocks

Among the many existing serial communication protocols, PCIe deserves special attention because of its role as high-speed solution for point-to-point processor communication. Due to this, FPGA vendors have been progressively including resources to support the implementation of PCIe buses, from the initial IP-based solutions to the currently available dedicated hardware blocks (Curd 2012).

From its initial definition (PCI-SIG 2015) to date, three PCIe specifications have been released (a fourth one is pending publication), whose characteristics are listed in Table 2.1.

Many FPGAs (e.g., Microsemi's SmartFusion2 and IGLOO2, Xilinx's from Series 5 on, Altera's from Arria II on) include dedicated hardware blocks to support Gen 1 and Gen 2 specifications, and the most advanced ones (e.g., Xilinx' Virtex-7 XT and HT, Altera's Stratix 10) also support Gen 3. The combination of these blocks with transceivers and, in some

TABLE 2.1

PCIe Base Specifications

PCI Spec Revision	Link Speed (GT/s)	Max Bandwidth[a] (Gb/s)	Encoding Scheme	Overhead (%)
Gen 1	2.5	2.0	8 B/10 B	20
Gen 2	5.0	4.0	8 B/10 B	20
Gen 3	8.0	7.88	128 B/130 B	1.5
Gen 4[b]	16.0	15.76	128 B/130 B	1.5

[a] Theoretical value. The actual one is lower because of packet overhead, among other factors.
[b] Publication pending.

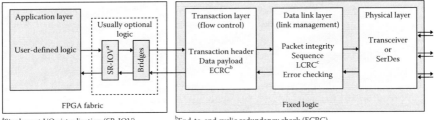

^aSingle root I/O virtualization (SR-IOV).   ^bEnd-to-end cyclic redundancy check (ECRC).
                                                 ^cLink cyclic redundancy check (LCRC).

FIGURE 2.23
Block diagram of a typical PCIe implementation.

cases, memory blocks allows the PCIe physical, data link, and transaction layers functions to be implemented (Figure 2.23), providing full endpoint and root-port functionality in ×1/×2/×4/×8/×16 lane configurations. The application layer is implemented in distributed logic. Communication with the transaction layer is achieved using interfaces usually based on AMBA buses.* A separate transceiver is needed for each lane, so the number of supported lanes depends on the availability of transceivers with PCIe capabilities.

In addition to basic specifications, some PCIe dedicated hardware blocks also support advanced functionalities, such as multiple-function, single-root I/O virtualization (SR-IOV), advanced error reporting (AER), and end-to-end CRC (ECRC).

The multiple-function feature allows several PCIe configuration header spaces to share the same PCIe link. From a software perspective, the situation is equivalent to having several PCIe devices, simplifying driver development (can be the same for all functions) and its portability.

The SR-IOV interface is an extension to the PCIe specification. When a single CPU (*single root*) runs different OSs (*multiple guests*) accessing an I/O device, SR-IOV can be used to assign a virtual configuration space to each OS, providing it with a direct link to the I/O device. In this way, data transfer rates can be very close to those achieved in a nonvirtualized solution.

AER and ECRC are optional functions oriented to systems with high reliability requirements. They allow detection, flagging, and correction of errors associated with PCIe links to be improved.

One of the major challenges for the implementation of PCIe is that, according to the Base Specification, links must be operational in less than 100 ms

---

* AMBA is a dominating de facto on-chip interconnect specification standard in industry for IP-based design (ARM-proprietary), which was first introduced in 1999 to ease the efficient interconnection of multiple processors and peripherals with different performances (low and high bandwidth). It is currently one of the most popular on-chip busing solutions for SoCs, and as such is analyzed in detail in Chapter 3.

after power on. Current FPGAs apply different configuration techniques to address this issue. One of them is partial reconfiguration (discussed in detail in Chapter 8): The FPGA is initially configured with a bitstream just containing the PCIe circuitry, and once it is operational, the rest of the FPGA functions required are configured on the fly using this link.

2.4.5 Serial Communication Interfaces

Although serial communication interfaces (such as I^2C, SPI, and USB) are usually required in many FPGA applications, not many devices include specialized hardware blocks with this kind of functionality, but it is implemented either using resources of the FPGA fabric or as part of an embedded hard or soft processor. At the moment, this book is being finalized, and to the best of authors' knowledge, only Lattice's and QuickLogic's devices include such hardware blocks. Lattice's MachXO2, MachXO3, iCE40LM, and iCE40 Ultra families as well as QuickLogic's ArcticLink II VX2 family include SPI and I^2C interfaces. USB and SD/SDIO/MMC/CE-ATA* interfaces are available in ArcticLink devices. Implementing such serial interfaces in hardware allows area, performance, and power consumption to be optimized.

The embedded function block (EFB) interface of the MachXO3 family (Lattice 2016) is shown in Figure 2.24a. It consists of a set of specialized hardware blocks, including one SPI and two I^2C interfaces. These three blocks are connected to the FPGA fabric through a Wishbone interface (analyzed in Section 3.5.4). The two I^2C interfaces can be configured as master (thus controlling other devices in the bus) or slave (thus acting as a resource available for a bus master). Among other features, they support 7 and 10 bit addressing, multimaster arbitration, interrupt request, and up to 400 kHz data transfer speed. The SPI block can also be configured as master or slave. It supports full-duplex data transfer, double-buffered data register, interrupt request, serial clock with programmable polarity and phase, and LSB- or MSB-first data transfer.

The iCE40 Ultra family (Lattice 2015), whose block diagram is shown in Figure 2.24b, includes up to two I^2C and two SPI interfaces, similar to those in the MachXO3 family. The distinct characteristic of iCE40 Ultra devices is that they can be categorized as "specific-purpose FPGAs," that is, configurable devices equipped with specific resources targeting specific applications rather than wide applicability (what most FPGAs are intended for). In this case, they are sensor managers targeting mobile platforms, such as smartphones, tablets, and handheld devices. With this purpose, in addition to the serial communication interfaces allowing them to connect to mobile sensors and application processors, they include other specialized hardware blocks, such as on-chip oscillators or DSP functional blocks.

---

* Secure Digital (SD), Secure Digital Input Output (SDIO), MultiMediaCard (MMC), and Consumer Electronic-ATA (CE-ATA) are memory card protocol definitions and standards used for solid-state storage.

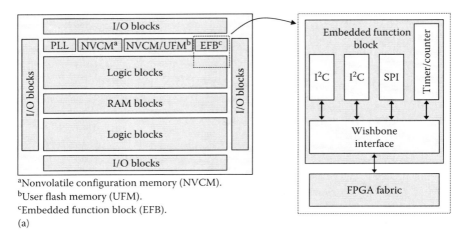

^aNonvolatile configuration memory (NVCM).
^bUser flash memory (UFM).
^cEmbedded function block (EFB).
(a)

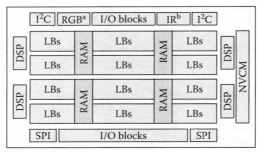

^aCurrent drive RGB LED outputs (RGB).
^bCurrent drive IR LED output (IR).
(b)

FIGURE 2.24
(a) MachXO3 EFB interface and (b) block diagram of iCE40 Ultra devices.

Similarly, QuickLogic's ArcticLink and ArcticLink II VX2 families are also oriented to mobile devices, so they include not only serial communication interfaces but also other very specific and complex blocks (only available in these devices and which are analyzed in Section 3.4.1). It is important to note that these FPGAs are nonvolatile devices based on QuickLogic proprietary ViaLink antifuse technology, and therefore one-time programmable (OTP), in contrast with the vast majority of FPGAs currently in the market, which are reconfigurable.

The block diagram of an ArcticLink II VX2 device (QuickLogic 2013) is shown in Figure 2.25a. It includes two serial interfaces: one SPI and one I²C. The I²C interface is mainly used as configuration bus for other embedded hardware blocks, although it can also be used as general-purpose interface. The SPI interface can only act as master, and it is intended for controlling

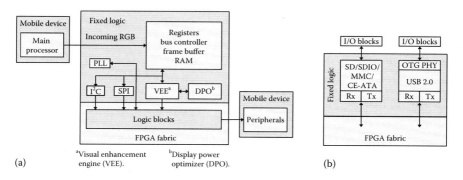

FIGURE 2.25
Block diagram of (a) ArcticLink II VX2 and (b) ArcticLink devices.

external elements such as sensors or displays. It supports up to three slaves and can operate in the frequency range from 1.5 to 27.2 MHz. These interfaces are not physically located in IOBs, but instead, they are connected by the user by means of resources of the FPGA fabric (see Figure 2.25a). This allows the number of external peripherals that can be connected to the interfaces to be extended by implementing a suitable multiplexing logic in the FPGA fabric.

Other resources included in ArcticLink devices (because they are widely used in handheld devices) are Hi-Speed USB 2.0 On-the-Go (OTG), and SD/SDIO/MMC/CE-ATA host controllers (Figure 2.25b) (QuickLogic 2010).

The Hi-Speed USB 2.0 OTG controller is a dual-role device supporting host and device functions. Its main features are as follows:

- Supports high- (480 Mbps), full- (12 Mbps), and low-speed (1.5 Mbps) transfers
- Integrated physical layer with dedicated internal PLL
- Supports both point-to-point and multipoint (root hub) applications
- Double-buffering scheme for improved throughput and data transfer capabilities
- Supports OTG Host Negotiation Protocol and Session Request Protocol
- Configurable power management features
- Integrated 5.2 kB FIFO
- Sixteen endpoints: one fixed bidirectional control endpoint, one software programmable IN or OUT endpoint, seven IN endpoints, and seven OUT endpoints

The SD/SDIO/MMC/CE-ATA controller is compliant with the *SD Host Controller Standard Specification, Version 2.0*. It supports clock rates up to 52 MHz; 1, 4, or 8 bit data modes; block size up to 512 bytes; and dynamic buffer management to increase data throughput.

References

Achronix. 2008. Introduction to Achronix FPGAs. White paper WP001-1.6.

Achronix. 2015. Speedster22i HD1000 FPGA data sheet DS005-1.0.

Actel (currently Microsemi). 2010. *ProASIC3 FPGA Fabric User's Guide.*

Altera. 2012. *Cyclone III Device Handbook.*

Altera. 2014. *Stratix V Device Handbook. Vol. 2: Transceivers.*

Altera. 2015a. MAX 10 FPGA device architecture.

Altera. 2015b. *Arria 10 Core Fabric and General Purpose I/Os Handbook.*

Altera. 2015c. *Stratix V Device Handbook. Vol. 1: Device Interfaces and Integration.*

Altera. 2015d. *Arria 10 Transceiver PHY User Guide UG-01143.*

Altera. 2016. *External Memory Interface Handbook Volume 1: Altera Memory Solution Overview, Design Flow, and General Information.*

Barrett, C. 1999. Fractional/integer-N PLL basics. Texas Instruments technical brief SWRA029. Texas Instruments, Dallas, TX.

Cortina Systems and Cisco Systems. 2008. Interlaken protocol definition. Revision 1.2.

Curd, D. 2012. PCI express for the 7 series FPGAs. Xilinx white paper WP384 (v1.1).

Gentile, K. 2008. Introduction to zero-delay clock timing techniques. Analog Devices application note AN-0983. Analog Devices, Norwood, MA.

Hutton, M. 2015. Understanding how the new HyperFlex architecture enables next-generation high-performance systems. Altera white paper WP-01231-1.0.

Jiao, B. 2015. Leveraging UltraScale FPGA transceivers for high-speed serial I/O connectivity. Xilinx white paper WP458 (v1.1).

Kuon, I., Tessier, R., and Rose, J. 2007. FPGA architecture: Survey and challenges. *Foundations and Trends in Electronic Design Automation*, 2:135–253.

Lattice. 2015. iCE40 Ultra family datasheet DS1048 (v1.8).

Lattice. 2016. MachXO3 family datasheet DS1047 (v1.6).

Microsemi. 2014. Fusion family of mixed signal FPGAs datasheet. Revision 6.

Microsemi. 2015a. IGLOO2 FPGA and SmartFusion2 SoC FPGA: Datasheet DS0451.

Microsemi. 2015b. SmartFusion2 SoC and IGLOO2 FPGA fabric: User guide UG0445.

Microsemi. 2015c. ProASIC3E flash family FPGAs: Datasheet DS0098.

Microsemi. 2015d. SmartFusion2 and IGLOO2 clocking resources: User guide UG0449.

PCI-SIG. 2015. PCI Express® base specification revision 3.1a. Available at: https://pcisig.com/specifications/pciexpress. Accessed November 20, 2016.

QuickLogic. 2010. ArcticLink solution platform datasheet (rev. M).

QuickLogic. 2013. ArcticLink II VX2 solution platform datasheet (rev. 1.0).

Rodriguez-Andina, J.J., Moure, M.J., and Valdes, M.D. 2007. Features, design tools, and application domains of FPGAs. *IEEE Transactions on Industrial Electronics*, 54:1810–1823.

Rodriguez-Andina, J.J., Valdes, M.D., and Moure, M.J. 2015. Advanced features and industrial applications of FPGAs—A review. *IEEE Transactions on Industrial Informatics*, 11:853–864.

Saban, K. 2012. Xilinx Stacked Silicon Interconnect Technology delivers breakthrough FPGA capacity, bandwidth, and power efficiency. Xilinx white paper WP380 (v1.2).

Texas Instruments. 2008. Fractional N frequency synthesis. Application note AN-1879.

Xilinx. 2004. Celebrating 20 years of innovation. *Xcell Journal*, 48:14–16.

Xilinx. 2006. Virtex-5 platform FPGA family technical backgrounder.

Xilinx. 2010. *Spartan-6 FPGA Configurable Logic Block: User Guide UG384 (v1.1).*

Xilinx. 2014a. *Spartan-6 FPGA SelectIO Resources: User Guide UG381 (v1.6).*

Xilinx. 2014b. *7 Series FPGAs Configurable Logic Block: User Guide UG474 (v1.7).*

Xilinx. 2014c. *7 Series FPGAs Memory Resources: User Guide UG473 (v1.11).*

Xilinx. 2014d. *7 Series FPGAs GTP Transceivers: User Guide UG482 (v1.8).*

Xilinx. 2015a. *7 Series FPGAs SelectIO Resources: User Guide UG471 (v1.5).*

Xilinx. 2015b. *7 Series FPGAs Clocking Resources: User Guide UG472 (v1.11.2).*

Xilinx. 2015c. *7 Series FPGAs GTX/GTH Transceivers: User Guide UG476 (v1.11).*

3

Embedded Processors in
FPGA Architectures

3.1 Introduction

Only 10 years ago we would have thought about the idea of a smart watch enabling us to communicate with a mobile phone, check our physical activity or heart rate, get weather forecast information, access a calendar, receive notifications, or give orders by voice as the subject of a futuristic movie. But, as we know now, smart watches are only one of the many affordable gadgets readily available in today's market.

The mass production of such consumer electronics devices providing many complex functionalities comes from the continuous evolution of electronic fabrication technologies, which allows SoCs to integrate more and more powerful processing and communication architectures in a single device, as shown by the example in Figure 3.1.

FPGAs have obviously also taken advantage of this technological evolution. Actually, the development of FPSoC solutions is one of the areas (if not THE area) FPGA vendors have concentrated most of their efforts on over recent years, rapidly moving from devices including one general-purpose microcontroller to the most recent ones, which integrate up to 10 complex processor cores operating concurrently. That is, there has been an evolution from FPGAs with *single-core* processors to *homogeneous* or *heterogeneous multicore* architectures (Kurisu 2015), with *symmetric multiprocessing* (SMP) or *asymmetric multiprocessing* (AMP) (Moyer 2013).

This chapter introduces the possibilities FPGAs currently offer in terms of FPSoC design, with different hardware/software alternatives. But, first of all, we will discuss the broader concept of SoC and introduce the related terminology, which is closely linked to processor architectures.

From Chapter 1, generically speaking, a SoC can be considered to consist of one or more programmable elements (general-purpose processors, microcontrollers, DSPs, FPGAs, or application-specific processors) connected to and interacting with a set of specialized peripherals to perform a set of tasks. From this concept, a single-core, single-thread processor (general-purpose,

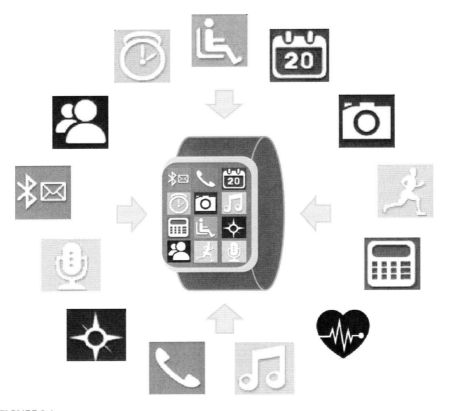

FIGURE 3.1
Processing and communication features in a smart watch SoC.

microcontroller, or DSP) connected to memory resources and specialized peripherals would usually be the best choice for embedded systems aimed at providing specific, non-time-critical functionalities. In these architectures, the processor acts as system master controlling data flows, although, in some cases, peripherals with memory access capabilities may take over data transfers with memory during some time intervals. Using FPGAs in this context provides higher flexibility than nonconfigurable solutions, because whenever a given software-implemented functionality does not provide good-enough timing performance, it can be migrated to hardware. In this solution, all hardware blocks are seen by the processor as peripherals connected to the same communication bus.

In order for single-core architectures to cope with continuous market demands for faster, more computationally powerful, and more energy-efficient solutions, the only option would be to increase operating frequency (taking advantage of nanometer-scale or 3D stacking technologies) and to reduce power consumption (by reducing power supply voltage). However, from the discussion in Chapter 1, it is clear that for the most demanding current applications, this is not a viable solution, and the only ones that may work are those based on the use of parallelism, that is, the ability of a system to execute several tasks concurrently.

The straightforward approach to parallelism is the use of multiple single-core processors (with the corresponding multiple sources of power consumption) and the distribution of tasks among them so that they can operate concurrently. In these architectures, memory resources and peripherals are usually shared among the processors and all elements are connected through a common communication bus. Another possible solution is the use of multithreading processors, which take advantage of dead times during the sequential execution of programs (for instance, while waiting for the response from a peripheral or during memory accesses) to launch a new thread executing a new task. Although this gives the impression of parallel execution, it is just in fact multithreading. Of course, these two relatively simple (at least conceptually) options are valid for a certain range of applications, but they have limited applicability, for instance, because of interconnection delays between processors or saturation of the multithreading capabilities.

3.1.1 Multicore Processors

The limitations of the aforementioned approaches can be overcome by using multicore processors, which integrate several processor cores (either multithreading or not) on a single chip. Since in most processing systems the main factor limiting performance is memory access time, trying to achieve improved performance by increasing operating frequency (and, hence, power consumption) does not make sense above certain limits, defined by the characteristics of the memories. Multicore systems are a much more efficient solution than that because they allow tasks to be executed concurrently by cores operating at lower frequencies than those a single processor would require, while reducing communication delays among processors, all of them within the same chip. Therefore, these architectures provide a better performance–power consumption trade-off.

3.1.1.1 Main Hardware Issues

There are many concepts associated with multicore architectures, and the commercial solutions to tackle them are very diverse. This section concentrates just on the main ideas allowing to understand and assess the ability of FPGAs to support SoCs. Readers can easily find additional information in the specialized literature about computer architecture (Stallings 2016).

The first multicore processors date back some 15 years ago, when IBM introduced the POWER4 architecture (Tendler et al. 2002). The evolution since then resulted in very powerful processing architectures, capable of supporting different OSs on a single chip. One might think the ability to integrate multiple cores would have a serious limitation related to increased silicon area

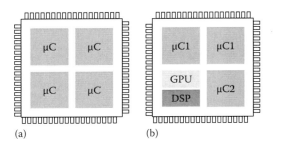

(a)   (b)

FIGURE 3.2
(a) Homogeneous and (b) heterogeneous multicore processor architectures. (a) Homogeneous architecture: processor cores are identical; (b) heterogeneous architecture: combines different processor cores.

and, in turn, cost. However, nanometer-scale and, more recently, 3D stacking technologies have enabled the fabrication of multicore chips at reasonably affordable prices. Today, one may easily find 16-core chips in the market.

As shown in Figure 3.2, multicore processors may be homogeneous (all of whose cores have the same architecture and instruction set) or heterogeneous (consisting of cores with different architectures and instruction sets). Most general-purpose multicore processors are homogeneous. In them, tasks (or threads) are interchangeable among processors (even at run time) with no effect on functionality, according to the availability of processing power in the different cores. Therefore, homogeneous solutions make an efficient use of parallelization capabilities and are easily scalable.

In spite of the good characteristics of homogeneous systems, there is a current trend toward heterogeneous solutions. This is mainly due to the very nature of the target applications, whose increasing complexity and growing need for the execution of highly specialized tasks require the use of platforms combining different architectures, as, for instance, microcontrollers, DSPs, and GPUs. Therefore, heterogeneous architectures are particularly suitable for applications where functionality can be clearly partitioned into specific tasks requiring specialized processors and not needing intensive communication among tasks.

Communications is actually a key aspect of any embedded system, but even more for multicore processors, which require low-latency, high-bandwidth communications not only between each processor and its memory/peripherals but also among the processors themselves. Shared buses may be used for this purpose, but the most current SoCs rely on crossbar interconnections (Vadja 2011). Given the importance of this topic, the on-chip buses most widely used in FPSoCs are analyzed in Section 3.5.

To reduce data traffic, multicore systems usually have one or two levels of local cache memory associated with each processor (so it can access the data it uses more often without affecting the other elements in the system),

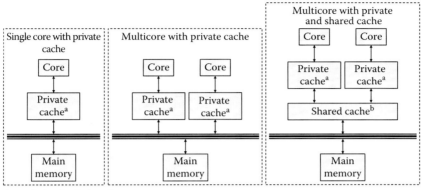

^aUsually level 1 (L1) cache.
^bUsually level 2 (L2) cache.

FIGURE 3.3
Usual cache memory architectures.

plus one higher level of shared cache memory. A side benefit of using shared memory is that in case the decision is made to migrate some sequential programming to concurrent hardware or vice versa, the fact that all cores share a common space reduces the need for modifications in data or control structures. Examples of usual cache memory architectures are shown in Figure 3.3.

The fact that some data (shared variables) can be modified by different cores, together with the use of local cache memories, implicitly creates problems related to data coherence and consistency. In brief, coherence means all cores see any shared variable as if there were no cache memories in the system, whereas consistency means instructions to access shared variables are programmed in the sharing cores in the right order. Therefore, coherence is an architectural issue (discussed in the following) and consistency a programming one (beyond the scope of this book).

A multicore system using cache memories is coherent if it ensures all processors sharing a given memory space always "see" at any position within it the last written value. In other words, a given memory space is coherent if a core reading a position within it retrieves data according to the order the cores sharing that variable have written values for it in their local caches. Coherence is obviously a fundamental requirement to ensure all processors access correct data at any time. This is the reason why all multicore processors include a cache-coherent memory system.

Although there are many different approaches to ensure coherence, all of them are based on modification–invalidation–update mechanisms. In a simplistic way, this means that when a core modifies the value of a shared variable in its local cache, copies of this variable in all other caches are invalidated and must be updated before they can be used.

3.1.1.2 Main Software Issues

As in the case of hardware, there are many software concepts to be considered in embedded systems, and multicore ones in particular, at different levels (application, middleware, OS) including, but not limited to, the necessary mechanisms for multithreading control, partitioning, resource sharing, or communications.

Different scenarios are possible depending on the complexity of the software to be executed by the processor and that of the processor itself,* as shown in Figure 3.4. For simple programs to be executed in low-end processors, the usual approach is to use bare-metal solutions, which do not require any software control layer (kernel or OS). Two intermediate cases are the implementation of complex applications in low-end processors or simple applications in high-end processors. In both cases, it is usual (and advisable) to use at least a simple kernel. Although this may not be deemed necessary for the latter case, it is highly recommended for the resulting system to be easily scalable. Finally, in order to efficiently implement complex applications in high-end processors, a real-time or high-end OS is necessary (Walls 2014). Currently, this is the case for most embedded systems.

Other important issues to be considered are the organization of shared resources, task partitioning and sequencing, as well as communications between tasks and between processors. From the point of view of the software architecture, these can be addressed by using either AMP or SMP approaches, depicted in Figure 3.5.

SMP architectures apply to homogeneous systems with two or more cores sharing memory space. They are based on using only one OS (if required) for

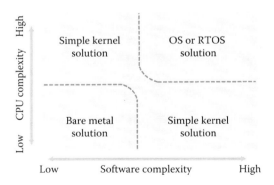

FIGURE 3.4
Software scenarios.

---

* Just to have a straightforward idea about complexity, we label as low-end processors those whose data buses are up to 16-bit wide and as high-end processors those with 32-bit or wider data buses.

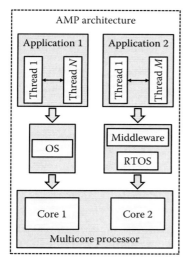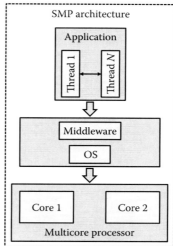

FIGURE 3.5
AMP and SMP multiprocessing.

all cores. Since the OS has all the information about the whole system hardware at any point, it can efficiently perform a dynamic distribution of the workload among cores (which implies extract application parallelism, partition of tasks/threads, and dynamic assignment of tasks to cores), as well as the control of the ordering of task completion and of resource sharing among cores. Resource sharing control is one of the most important advantages of SMP architectures. Another significant one is easy interprocess communication, because there is no need for implementing any specific communication protocol, thus avoiding the overhead this would introduce. Finally, debugging tasks are simpler when working with just one OS.

SMP architectures are clearly oriented to get the most possible advantage of parallelism to maximize processing performance, but they have a main limiting factor, related to the dynamic distribution of workload. This factor affects the ability of the system to provide a predictable timing response, which is a fundamental feature in many embedded applications. Another past drawback, the need for an OS supporting multicore processing, is not a significant problem anymore given the wide range of options currently available (Linux, embedded Windows, and Android, to cite just some).

In contrast to SMP, AMP architectures can be implemented in either homogeneous or heterogeneous multicore processors. In this case, each core runs its own OS (either separate copies of the same or totally different ones; some cores may even implement a bare-metal system). Since none of the OSs is specifically in charge of controlling shared resources, such control must be very carefully performed at the application level. AMP solutions are oriented to applications with a high level of intrinsic parallelism, where critical tasks are assigned to specific resources in order for a predictable behavior to be achieved. Usually, in AMP systems, processes are *locked* (assigned) to a given processor. This simplifies the individual control of each core by the designer. In addition, it eases migration from single-core solutions.

3.1.2 Many-Core Processors

Single- and multicore solutions are the ones most commonly found in SoCs, but there is a third option, many-core processors, which find their main niche in systems requiring a high scalability (mostly intensive computing applications), for instance, cloud computing datacenters. Many-core processors consist of a very large number of cores (up to 400 in some existing commercially available solutions [Nickolls and Dally 2010; NVIDIA 2010; Kalray 2014]), but are simpler and have less computing power than those used in multicore systems. These architectures aim at providing massive concurrency with a comparatively low energy consumption. Although many researchers and vendors (Shalf et al. 2009; Jeffers and Reinders 2015; Pavlo 2015) claim this will be the dominant processing architecture in the future, its analysis is out of the scope of this book, because until now, it has not been adopted in any FPGA.

3.1.3 FPSoCs

At this point two pertinent questions arise: What is the role of FPGAs in SoC design, and what can they offer in this context? Obviously, when speaking of parallelism or versatility, no hardware platform compares to FPGAs. Therefore, combining FPGAs with microcontrollers, DSPs, or GPUs clearly seems to be an advantageous design alternative for a countless number of applications demanded by the market. Some years ago, FPGA vendors realized the tremendous potential of SoCs and started developing chips that combined FPGA fabric with embedded microcontrollers, giving rise to FPSoCs.

The evolution of FPSoCs can be summarized as shown in Figure 3.6. Initially, FPSoCs were based on single-core soft processors, that is, configurable microcontrollers implemented using the logic resources of the FPGA fabric. The next step was the integration in the same chip as the FPGA fabric of single-core hard processors, such as PowerPC. In the last few years, several families of FPGA devices have been developed that integrate multicore processors (initially homogeneous architectures and, more recently, heterogeneous ones). As a result, the FPSoC market now offers a wide portfolio of low-cost, mid-range, and high-end devices for designers to choose from depending on the performance level demanded by the target application.

FPGAs are among the few types of devices that can take advantage of the latest nanometer-scale fabrication technologies. At the time of writing this book, according to FPGA vendors (Xilinx 2015; Kenny 2016), high-end FPGAs are fabricated in 14 nm process technologies, but new families have already

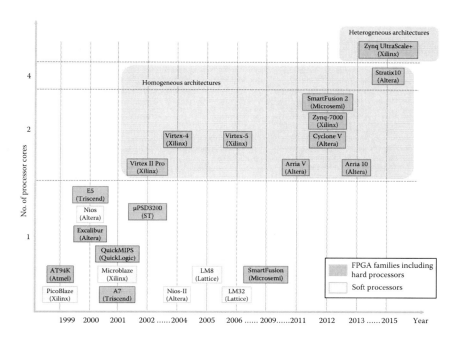

FIGURE 3.6
FPSoC evolution.

been announced based on 10 nm technologies, whereas the average for ASICs is 65 nm. The reason for this is just economic viability. When migrating a chip design to a more advanced node (let us say from 28 to 14 nm), the costs associated with hardware and software design and verification dramatically grow, to the extent that for the migration to be economically viable, the return on investment must be in the order of hundreds of millions of dollars. Only chips for high-volume applications or those that can be used in many different applications (such as FPGAs) can get to those figures.

The different FPSoC options currently available in the market are analyzed in the following sections.

3.2 Soft Processors

As stated in Section 3.1.3, soft processors are involved in the origin of FPSoC architectures. They are processor IP cores (usually general-purpose ones) implemented using the logic resources of the FPGA fabric (distributed logic, specialized hardware blocks, and interconnect resources), with the advantage of having a very flexible architecture.

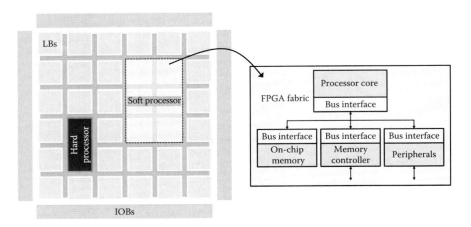

FIGURE 3.7
Soft processor architecture.

As shown in Figure 3.7, a soft processor consists of a processor core, a set of on-chip peripherals, on-chip memory, and interfaces to off-chip memory. Like microcontroller families, each soft processor family uses a consistent instruction set and programming model.

Although some of the characteristics of a given soft processor are pre-defined and cannot be modified (e.g., the number of instruction and data bits, instruction set architecture [ISA], or some functional blocks), others can be defined by the designer (e.g., type and number of peripherals or memory map). In this way, the soft processor can, to a certain extent, be tailored to the target application. In addition, if a peripheral is required that is not available as part of the standard configuration possibilities of the soft processor, or a given available functionality needs to be optimized (for instance, because of the need to increase processing speed in performance-critical systems), it is always possible for the designer to implement a custom peripheral using available FPGA resources and connect it to the CPU in the same way as any "standard" peripheral.

The main alternative to soft processors are hard processors, which are fixed hardware blocks implementing specific processors, such as the ARM's Cortex-A9 (ARM 2012) included by Altera and Xilinx in their latest fami-lies of devices. Although hard processors (analyzed in detail in Section 3.3) provide some advantages with regard to soft ones, their fixed architecture causes not all their resources to be necessary in many applications, whereas in other cases there may not be enough of them. Flexibility then becomes the main advantage of soft processors, enabling the development of custom solu-tions to meet performance, complexity, or cost requirements. Scalability and reduced risk of obsolescence are other significant advantages of soft proces-sors. Scalability refers to both the ability of adding resources to support new features or update existing ones along the whole lifetime of the system and the possibility of replicating a system, implementing more than one processor in the same FPGA chip. In terms of reduced risk of obsolescence, soft processors can usually be migrated to new families of devices. Limiting factors in this regard are that the soft processor may use logic resources specific to a given family of devices, which may not be available in others, or that the designer is not the actual owner of the HDL code describing the soft processor.

Soft processor cores can be divided into two groups:

1. Proprietary cores, associated with an FPGA vendor, that is, supported only by devices from that vendor.
2. Open-source cores, which are technology independent and can, therefore, be implemented in devices from different vendors.

These two types of soft processors are analyzed in Sections 3.2.1 and 3.2.2, respectively. Although there are many soft processors with many diverse features available in the market, without loss of generality, we will focus on the main features and the most widely used cores, which will give a fairly comprehensive view of the different options available for designers.

3.2.1 Proprietary Cores

Proprietary cores are optimized for a particular FPGA architecture, so they usually provide a more reliable performance, in the sense that the information about processing speed, resource utilization, and power consumption can be accurately determined, because it is possible to simulate their behavior from accurate hardware models. Their major drawback is that the portability of and the possibility of reusing the code are quite limited.

Open-source cores are portable and more affordable. They are relatively easy to adapt to different FPGA architectures and to modify. On the other hand, not being optimized for any particular architecture, usually, their performance is worse and less predictable, and their implementation requires more FPGA resources to be used.

Xilinx's PicoBlaze (Xilinx 2011a) and MicroBlaze (Xilinx 2016a) and Altera's Nios-II* (Altera 2015c), whose block diagrams are shown in Figure 3.8a through c, respectively, have consistently been the most popular proprietary processor cores over the years. More recently, Lattice Semiconductor released the LatticeMico32 (LM32) (Lattice 2012) and LatticeMico8 (LM8) (Lattice 2014) processors,[†] whose block diagrams are shown in Figure 3.8d and e, respectively.

---

* Altera previously developed and commercialized the Nios soft processor, predecessor of Nios-II.
[†] Although LM8 and LM32 are actually open-source, free IP cores, since they are optimized for Lattice FPGAs, they are better analyzed together with proprietary cores.

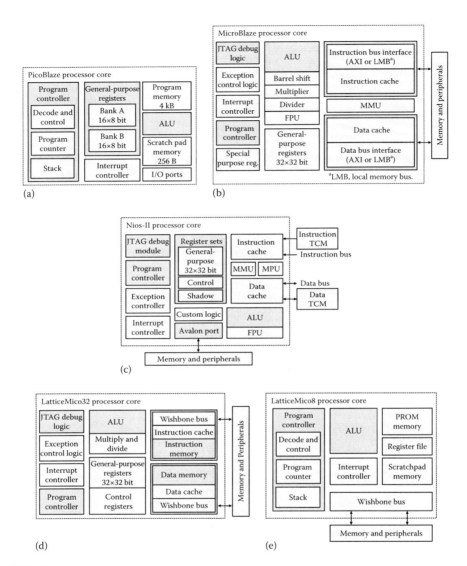

FIGURE 3.8
Block diagrams of proprietary processor cores: (a) Xilinx's PicoBlaze, (b) Xilinx's MicroBlaze, (c) Altera's Nios-II, (d) Lattice's LM32, and (e) Lattice's LM8.

PicoBlaze and LM8 are 8-bit RISC microcontroller cores optimized for Xilinx* and Lattice FPGAs, respectively. Both have a predictive behavior, particularly PicoBlaze, all of whose instructions are executed in two clock cycles. Both have also similar architectures, including:

---

* KCPSM3 is the PicoBlaze version for Spartan-3 FPGAs, and KCPSM6 for Spartan-6, Virtex-6, and Virtex-7 Series.

- General-purpose registers (16 in PicoBlaze, 16 or 32 in LM8).
- Up to 4 K of 188-bit-wide instruction memory.
- Internal scratchpad RAM memory (64 bytes in PicoBlaze, up to 4 GB in 256-byte pages in LM8).
- Arithmetic Logic Unit (ALU).
- Interrupt management (one interrupt source in PicoBlaze, up to 8 in LM8).

The main difference between PicoBlaze and LM8 is the communication interface. None of it includes internal peripherals, so all required peripherals must be separately implemented in the FPGA fabric. PicoBlaze communicates with them through up to 256 input and up to 256 output ports, whereas LM8 uses a Wishbone interface from OpenCores, described in Section 3.5.4.

Similarly, although MicroBlaze, Nios-II, and LM32 are also associated with the FPGAs of their respective vendors, they have many common characteristics and features:

- 32-bit general-purpose RISC processors.
- 32-bit instruction set, data path, and address space.
- Harvard architecture.
- Thirty-two 32-bit general-purpose registers.
- Instruction and data cache memories.
- Memory management unit (MMU) to support OSs requiring virtual memory management (only in MicroBlaze and Nios-II).
- Possibility of variable pipeline, to optimize area or performance.
- Wide range of standard peripherals such as timers, serial communication interfaces, general-purpose I/O, SDRAM controllers, and other memory interfaces.
- Single-precision floating point computation capabilities (only in MicroBlaze and Nios-II).
- Interfaces to off-chip memories and peripherals.
- Multiple interrupt sources.
- Exception handling capabilities.
- Possibility for creating and adding custom peripherals.
- Hardware debug logic.
- Standard and real-time OS support: Linux, µCLinux, MicroC/OS-II, ThreadX, eCos, FreeRTOS, uC/OS-II, or embOS (only in MicroBlaze and Nios-II).

A soft processor is designed to support a certain ISA. This implies the need for a set of functional blocks, in addition to instruction and data memories, peripherals, and resources, to connect the core to external elements. The functional blocks supporting the ISA are usually implemented in hardware, but some of them can also be emulated in software to reduce FPGA resource usage. On the other hand, not all blocks building up the core are required for all applications. Some of them are optional, and it is up to the designer whether to include them or not, according to system requirements for functionality, performance, or complexity. In other words, a soft processor core does not have a fixed structure, but it can be adapted to some extent to the specific needs of the target application.

Most of the remainder of this section is focused on the architecture of the Nios-II soft processor core as an example, but a vast majority of the analyses are also applicable to any other similar soft processors. As shown in Figure 3.8c, the Nios-II architecture consists of the following functional blocks:

- *Register sets*: They are organized in thirty-two 32-bit general-purpose registers and up to thirty-two 32-bit control registers. Optionally, up to 63 shadow register sets may be defined to reduce context switch latency and, in turn, execution time.

- *ALU*: It operates with the contents of the general-purpose registers and supports arithmetic, logic, relational, and shift and rotate instructions. When configuring the core, designers may choose to have some instructions (e.g., division) implemented in hardware or emulated in software, to save FPGA resources for other purposes at the expense of performance.

- *Custom instruction logic (optional)*: Nios-II supports the addition of not only custom components but also of custom instructions, for example, to accelerate algorithm execution. The idea is for the designer to be able to substitute a sequence of native instructions by a single one executed in hardware. Each new custom instruction created generates a logic block that is integrated in the ALU, as shown in Figure 3.9. This is an interesting feature of the Nios-II architecture not provided by others.

  Up to 256 custom instructions of five different types (combinational, multicycle, extended, internal register file, and external interface) can be supported. A combinational instruction is implemented through a logic block that performs its function within a single clock cycle, whereas multicycle (sequential) instructions require more than one clock cycle to be completed. Extended instructions allow several (up to 256) combinational or multicycle instructions to be implemented in a single logic block. Internal register file custom instructions are those that can operate with the internal registers of their logic block instead of with Nios-II general-purpose registers (the ones used by other custom instructions and by native instructions).

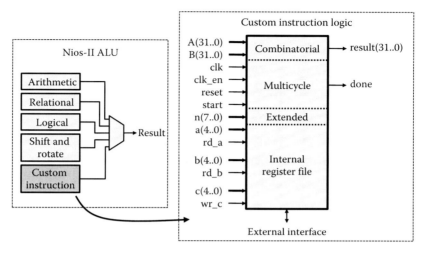

FIGURE 3.9
Connection of custom instruction logic to the ALU.

Finally, external interface custom instructions generate communication interfaces to access elements outside of the processor's data path.

Whenever a new custom instruction is created, a macro is generated that can be directly instantiated in any C or C++ application code, eliminating the need for programmers to use assembly code (they may use it anyway if they wish) to take advantage of custom instructions.

In addition to user-defined instructions, Nios-II offers a set of predefined instructions built from custom instruction logic. These include single-precision floating-point instructions (according to IEEE Std. 754-2008 or IEEE Std. 754-1985 specifications) to support computation-intensive floating-point applications:

- *Exception controller*: It provides response to all possible exceptions, including internal hardware interrupts, through an exception handler that assesses the cause of the exception and calls the corresponding exception response routine.

- *Internal and external interrupt controller* (*EIC*) (*optional*): Nios-II supports up to 32 internal hardware interrupt sources, whose priority is determined by software. Designers may also create an EIC and connect it to the core through an EIC interface. When using EIC, internal interrupt sources are also connected to it and the internal interrupt controller is not implemented.

- *Instruction and data buses*: Nios-II is based on a Harvard architecture. The separate instruction and data buses are both implemented using 32-bit Avalon-MM master ports, according to Altera's proprietary Avalon interface specification. The Avalon bus is analyzed in Section 3.5.2.

The data bus allows memory-mapped read/write access to both data memory and peripherals, whereas the instruction bus just fetches (reads) the instructions to be executed by the processor. Nios-II architecture does not specify the number or type of memories and peripherals that can be used, nor the way to connect to them either. These features are configured when defining the FPSoC. However, most usually, a combination of (fast) on-chip embedded memory, slower off-chip memory, and on-chip peripherals (implemented in the FPGA fabric) is used.

- *Instruction and data cache memories* (*optional*): Cache memories are supported in the instruction and data master ports. Both instruction and data caches are an intrinsic part of the core, but their use is optional. Software methods are available to bypass one of them or both. Cache management and coherence are managed in software.

- *Tightly coupled memories (TCM)* (*optional*): The Nios-II architecture includes optional TCM ports aimed at ensuring low-latency memory access in time-critical applications. These ports connect both instruction and data TCMs, which are on chip but external to the core. Several TCMs may be used, each one associated with a TCM port.

- *MMU* (*optional*): This block handles virtual memory, and, therefore, its use makes only sense in conjunction with an OS requiring virtual memory. Its main tasks are memory allocation to processes, translation of virtual (software) memory addresses into physical addresses (the ones the hardware sets in the address lines of the Avalon bus), and memory protection to prevent any process to write to memory sections without proper authorization, thus avoiding errant software execution.

- *Memory protection unit (MPU)* (*optional*): This block is used when memory protection features are required but virtual memory management is not. It allows access permissions to the different regions in the memory map to be defined by software. In case a process attempts to perform an unauthorized memory access, an exception is generated.

- *JTAG debug module* (*optional*): As shown in Figure 3.10, this block connects to the on-chip JTAG circuitry and to internal core signals. This allows the soft processor to be remotely accessed for debugging purposes. Some of the supported debugging tasks are downloading programs to memory, starting and stopping program execution, setting breakpoints and watchpoints, analyzing and editing registers and memory contents, and collecting real-time execution trace data. In this context, the advantage with regard to hard processors is that the debugging module can be used during the design and verification phase and removed for normal operation, thus releasing FPGA resources.

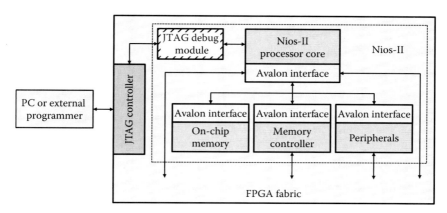

FIGURE 3.10
Connection of the JTAG debug module.

To ease the task of configuring the Nios-II architecture to fit the requirements of different applications, Altera provides three basic models from which designers can build their own core, depending on whether performance or complexity weighs more in their decisions. Nios-II/f (fast) is designed to maximize performance at the expense of FPGA resource usage. Nios-II/s (standard) offers a balanced trade-off between performance and resource usage. Finally, Nios-II/e (economy) optimizes resource usage at the expense of performance.

The similarities between the hardware architecture of Altera's Nios-II and Xilinx MicroBlaze can be clearly noticed in Figure 3.8. Both are 32-bit RISC processors with Harvard architecture and include fixed and optional blocks, most of which are present in the two architectures, even if there may be some differences in the implementation details. Lattice's LM32 is also a 32-bit RISC processor, but much simpler than the two former ones. For instance, it does not include an MMU block. It can be integrated with OSs such as μCLinux, uC/OS-II, and TOPPERS/JSP kernel (Lattice 2008).

The core processor is not the only element a soft processor consists of, but it is the most important one, since it has to ensure that any instruction in the ISA can be executed no matter what the configuration of the core is. In addition, the soft processor includes peripherals, memory resources, and the required interconnections. A large number of peripherals are or may be integrated in the soft processor architecture. They range from standard resources (GPIO, timers, counters, or UARTs) to complex, specialized blocks oriented to signal processing, networking, or biometrics, among other fields. Not only FPGA vendors provide peripherals to support their soft processors, but others are also available from third parties.

Communication of the core processor with peripherals and external circuits in the FPGA fabric is a key aspect in the architecture of soft processors. In this regard, there are significant differences among the three soft processors being analyzed. Nios-II has always used, from its very first versions to date, Altera's proprietary Avalon bus. On the other hand, Xilinx initially used IBM's CoreConnect bus, together with proprietary ones (such as local memory bus [LMB] and Xilinx CacheLink [XCL]), but the most current devices use ARM's AXI interface. Lattice LM32 processor uses WishBone interfaces. A detailed analysis of the on-chip buses most widely used in FPSoCs is made in Section 3.5.

At this point, readers may be afraid to realize the huge amount and diversity of concepts, terms, hardware and software alternatives, or design decisions one must face when dealing with soft processors. Fortunately, designers have at their disposal robust design environments as well as an ecosystem of design tools and IP cores that dramatically simplify the design process. The tools supporting the design of SoPCs are described in Section 6.3.

3.2.2 Open-Source Cores

In addition to proprietary cores, associated with certain FPGA architectures/vendors, there are also open-source soft processor cores available from other parties. Some examples are ARM's Cortex-M1 and Cortex-M3, Freescale's ColdFire V1, MIPS Technologies' MP32, OpenRISC 1200 from OpenCores community, Aeroflex Gaisler's LEON4, as well as implementations of many different well-known processors, such as the 8051, 80186 (88), and 68000. The main advantages of these solutions are that they are technology independent, low cost, based on well-known, proven architectures, and they are supported by a full set of tools and OSs.

The Cortex-M1 processor (ARM 2008), whose block diagram is shown in Figure 3.11a, was developed by ARM specifically targeting FPGAs. It has a 32-bit RISC architecture and, among other features, includes configurable instruction and data TCMs, interrupt controller, or configurable debug logic. The communication interface is ARM's proprietary AMBA AHB-Lite 32-bit bus (described in Section 3.5.1.1). The core supports Altera, Microsemi, and Xilinx devices, and it can operate in a frequency range from 70 to 200 MHz, depending on the FPGA family.

The OpenRISC 1200 processor (OpenCores 2011) is based on the OpenRISC 1000 architecture, developed by OpenCores targeting the implementation of 32- and 64-bit processors. OpenRISC 1200, whose block diagram is shown in Figure 3.11b, is a 32-bit RISC processor with Harvard architecture. Among other features, it includes general-purpose registers, instructions and data caches, MMU, floating-point unit (FPU), MAC unit for the efficient implementation of signal processing functions, and exception/interrupt management units. The communication interface is WishBone (described in Section 3.5.4). It supports different OSs, such as Linux, RTEMS, FreeRTOS, and eCos.

LEON4 is a 32-bit processor based on the SPARC V8 architecture originated from European Space Agency's project LEON. It is one of the most complex and flexible (configurable) open-source cores. It includes an ALU

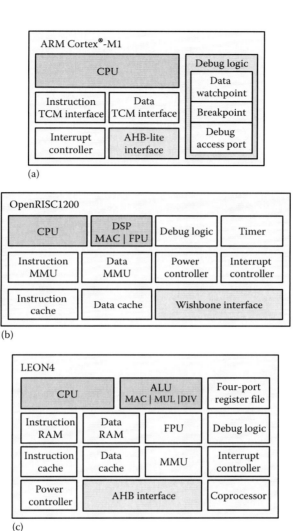

FIGURE 3.11
Some open-source soft processors: (a) Cortex-M1, (b) OpenRISC1200, and (c) LEON4.

with hardware multiply, divide, and MAC units, IEEE-754 FPU, MMU, and debug module with instruction and data trace buffer. It supports two levels of instruction and data caches and uses the AMBA 2.0 AHB bus (described in Section 3.5.1.1) as communication interface. From a software point of view, it supports Linux, eCos, RTEMS, Nucleus, VxWorks, and ThreadX.

Table 3.1 summarizes the performance of the different soft processors analyzed in this chapter. It should be noted that data have been extracted

TABLE 3.1

Performance of Soft Processors

Soft Processor	MIPS or DMIPS/MHz	Maximum Frequency Reported (MHz)
PicoBlaze	100 MIPS[a]	240
LatticeMico8	No data	94.6 (LatticeECP2)
MicroBlaze	1.34 DMIPS/MHz	343
Nios-II		
Nios-II/e	0.15 DMIPS/MHz	200
Nios-II/s	0.74 DMIPS/MHz	165
Nios-II/f	1.16 DMIPS/MHz	185
LatticeMico32	1.14 DMIPS/MHz	115
Cortex-M1	0.8 DMIPS/MHz	200
OpenRISC1200	1 DMIPS/MHz	300
LEON4	1.7 DMIPS/MHz	150

[a] Up to 200 MHz or 100 MIPS in a Virtex-II Pro FPGA (Xilinx 2011a).

from information provided by vendors and, in some cases, it is not clear how this information has been obtained.

Since several soft processors can be instantiated in an FPGA design (to the extent that there are enough resources available), many diverse FPSoC solutions can be developed based on them, from single to multicore. These multicore systems may be based on the same or different soft processors, or their combination with hard processors, and support different OSs. Therefore, it is possible to design homogeneous or heterogeneous FPSoCs, with SMP or AMP architectures.

3.3 Hard Processors

Soft processors are a very suitable alternative for the development of FPSoCs, but when the highest possible performance is required, hard processors may be the only viable solution. Hard processors are commercial, usually proprietary, processors that are integrated with the FPGA fabric in the same chip, so they can be somehow considered as another type of specialized hardware blocks. The main difference with the stand-alone versions of the same processors is that hard ones are adapted to the architectures of the FPGA devices they are embedded in so that they can be connected to the FPGA fabric with minimum delay. However, very interestingly, from the point of view of software developers, there is no difference, for example, in terms of architecture or ISA.

There are obviously many advantages derived from the use of optimized, state-of-the-art processors. Their performance is similar to the corresponding ASIC implementations (and well known from these implementations); they have a wide variety of peripherals and memory management resources, are highly reliable, and have been carefully designed to provide a good performance/functionality/power consumption trade-off. Documentation is usually extensive and detailed, and they have whole development and support ecosystems provided by the vendors. There are also usually many reference designs available that designers can use as starting point to develop their own applications.

Hard processors have also some drawbacks. First, they are not scalable, because their fixed hardware structure cannot be modified. Second, since they are fine-tuned for each specific FPGA family, design portability may be limited. Finally, same as for stand-alone processors, obsolescence affects hard processors. This is a market segment where new devices with ever-enhanced features are continuously being released, and as a consequence, production of (and support for) relatively recent devices may be discontinued.

The first commercial FPSoCs including hard processors were proposed by Atmel and Triscend.* For instance, Atmel developed the AT94K Field Programmable System Level Integrated Circuit series (Atmel 2002), which combined a proprietary 8-bit RISC AVR processor (1 MIPS/MHz, up to 25 MHz) with reconfigurable logic based on its AT40K FPGA family. Triscend, on its side, commercialized the E5 series (Triscend 2000), including an 8032 microcontroller (8051/52 compatible, 10 MIPS at 40 MHz). In both cases, the reconfigurable part consisted of resources accounting for roughly 40,000 equivalent logic gates, and the peripherals of the microcontrollers consisted of just a small set of timers/counters, serial communication interfaces (SPI, UART), capture and compare units (capable of generating PWM signals), and interrupt controllers (capable of handling both internal and external interrupt sources). None of these devices is currently available in the market, although Atmel still produces AT40K FPGAs.

After only a few months, 32-bit processors entered the FPSoC market with the release of Altera's Excalibur family (Altera 2002), followed by QuickLogic's QuickMIPS ESP (QuickLogic 2001), Triscend's A7 (Triscend 2001), and Xilinx's Virtex-II Pro (Xilinx 2011b), Virtex-4 FX (Xilinx 2008), and Virtex-5 FXT (Xilinx 2010b). This was a big jump ahead in terms of processor architectures, available peripherals, and operating frequencies/performance.

Altera and Triscend already opted at this point to include ARM processors in their FPSoCs, whereas QuickLogic devices combined a MIPS32 4Kc processor from MIPS Technologies† with approximately 550,000 equivalent logic gates of Via-Link fabric (QuickLogic's antifuse proprietary technology).

---

* Microchip Technology acquired Atmel in 2016, and Xilinx acquired Triscend in 2004.
† Imagination Technologies acquired MIPS Technologies in 2013.

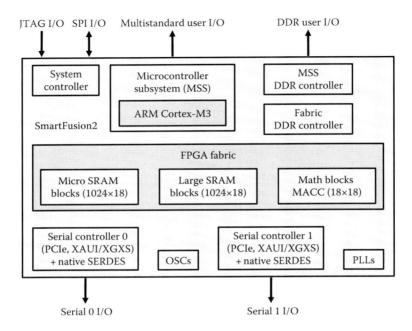

FIGURE 3.12
SmartFusion architecture.

Xilinx Virtex-II Pro and Virtex-4 FX devices included up to two IBM PowerPC 405 cores and Virtex-5 FX up to two IBM PowerPC 440 cores. Only these three latter families are still in the market, although Xilinx recommends not to use them for new designs.

It took more than 5 years for a new FPSoC family (Microsemi's SmartFusion, Figure 3.12) to be released, but since then there has been a tremendous evolution, with one common factor: All FPGA vendors opted for ARM architectures as the main processors for their FPSoC platforms. Microsemi's SmartFusion and SmartFusion 2 (Microsemi 2013, 2016) families include an ARM Cortex-M3 32-bit RISC processor (Harvard architecture, up to 166 MHz, 1.25 DMIPS/MHz), with two 32 kB SRAM memory blocks, 512 kB of 64-bit nonvolatile memory, and 8 kB instruction cache. It provides different interfaces (all based on ARM's proprietary AMBA bus, described in Section 3.5.1) for communication with specialized hardware blocks or custom user logic in the FPGA fabric, as well as many peripherals to support different communication standards (USB controller; SPI, I²C, and CAN blocks, multi-mode UARTs, or Triple-Speed Ethernet media access control). In addition, it includes an embedded trace macrocell block intended to ease system debug and setup.

Altera and Xilinx include ARM Cortex-A9 cores in some of their most recent FPGA families, such as Altera's Arria 10 (Figure 3.13), Arria V, and

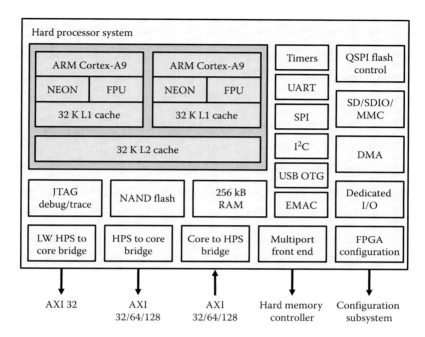

FIGURE 3.13
Arria 10 hard processor system.

Cyclone V (Altera 2016a,b) and Xilinx's Zynq-7000 AP SoC (Xilinx 2014). The ARM Cortex-A9 is a 32-bit dual-core processor (2.5 DMIPS/MHz, up to 1.5 GHz). Dual-core architectures are particularly suitable for real-time operation, because one of the cores may run the OS and main application programs, whereas the other core is in charge of time-critical (real-time) functions. In both Altera and Xilinx devices, the processors and the FPGA fabric are supplied from separate power sources. If only the processor is to be used, it is possible to turn off power supply for the fabric, hence allowing power consumption to be reduced. In addition, the logic can be fully or partially configured from the processor at any time.*

The main features of the ARM Cortex-A9 dual-core processor are as follows:

- Ability to operate in single-processor, SMP dual-processor, or AMP dual-processor modes.
- Up to 256 kB of on-chip RAM and 64 kB of on-chip ROM.
- Each core has its own level 1 (L1) separate instruction and data caches, 32 kB each, and share 512 kB of level 2 (L2) cache.

---

* FPGA configuration is analyzed in Chapter 6.

- Dynamic length pipeline (8–11 stages).
- Eight-channel DMA controller supporting different data transfer types: memory to memory, memory to peripheral, peripheral to memory, and scatter–gather.
- MMU.
- Single- and double-precision FPU.
- NEON media processing engine, which enhances FPU features by providing a quad-MAC and additional 64-bit and 128-bit register sets supporting single-instruction, multiple-data (SIMD) and vector floating-point instructions. NEON technology can accelerate multimedia and signal processing algorithms such as video encode/decode, 2D/3D graphics, gaming, audio and speech processing, image processing, telephony, and sound synthesis.
- Available peripherals include interrupt controller, timers, GPIO, or 10/100/1000 tri-mode Ethernet Media Access Control, as well as USB 2.0, CAN, SPI, UART, and I²C interfaces.
- Hard memory interfaces for DDR4, DDR3, DDR3L, DDR2, LPDDR2, flash (QSPI, NOR, and NAND), and SD/SDIO/MMC memories.
- Connections with the FPGA fabric (distributed logic and specialized hardware blocks) through AXI interfaces (described in Section 3.5.1.3).
- ARM CoreSight Program Trace Macrocell, which allows the instruction flow being executed to be accessed for debugging purposes (Sharma 2014).

At the time of writing of this book, the two most recently released FPSoC platforms are Altera's Stratix 10 (Altera 2015d) and Xilinx's Zynq UltraScale+ (Xilinx 2016b), both including an ARM Cortex-A53 quad-core processor. Most of the features of the ARM Cortex-A53 processor (Figure 3.14) are already present in the ARM Cortex-A9, but the former is smaller and has less power consumption. The cores in Stratix 10 and Zynq UltraScale+ families can operate up to 1.5 GHz, providing 2.3 DMIPS/MHz performance.

In addition, Zynq UltraScale+ devices include an ARM Cortex-R5 dual-core processor and an ARM Mali-400 MP2 GPU, as shown in Figure 3.14b, resulting in a heterogeneous multiprocessor SoC (MPSoC) hardware architecture. The ARM Cortex-R5 is a 32-bit dual-core real-time processor,* capable of operating at up to 600 MHz and providing 1.67 DMIPS/MHz performance. Cores can work in split (independent) or lock-step (parallel) modes.

---

* Cortex-A series includes "Application" processors and Cortex-R series "real-time" ones.

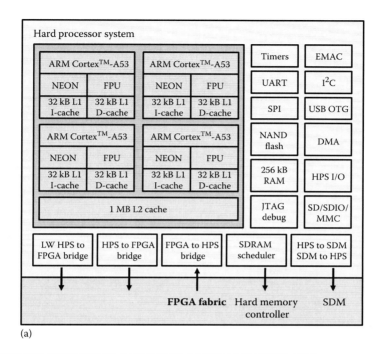

(a)

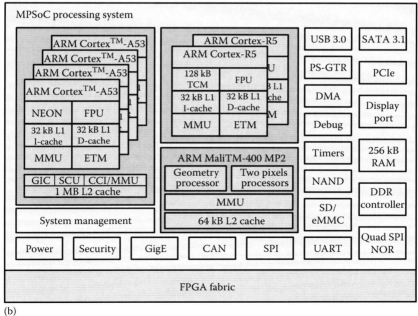

(b)

FIGURE 3.14
Processing systems of (a) Altera's Stratix 10 FPGAs and (b) Xilinx's Zynq UltraScale+ MPSoCs.

Lock-step operation is intended for safety-critical applications requiring redundant systems.

The main features of each core are as follows:

- 32 kB L1 instruction and data caches and 128 kB TCM for highly deterministic or low-latency applications (real-time single-cycle access). All memories have ECC and/or parity protection.
- Interrupt controller.
- MPU.
- Single- and double-precision FPU.
- Embedded trace macrocell for connection to ARM CoreSight debugging system.
- AXI interfaces (described in Section 3.5.1.3).

The ARM Mali-400 GPU is a low-power graphics acceleration processor, capable of operating at up to 667 MHz. Its 2D vector graphics are based on OpenVG 1.1,* whereas 3D graphics are based on OpenGL ES 2.0.† It supports Full Scene Anti-Aliasing and Ericsson Texture Compression to reduce external memory bandwidth and is fully autonomous to operate in parallel with the ARM Cortes-A53 application processor. It consists of five main blocks:

1. *Geometry processor, in charge of the vertex processing stage of the graphics pipeline*: It generates lists of primitives and accelerates building of data structures for pixel processors.
2. *Pixel processors (two), which handle the rasterization and fragment processing stages of the graphics pipeline*: They produce the framebuffer results that screens display as final images.
3. *MMU*: Both the geometry processor and the pixel processors use MMUs for access checking and translation.
4. *L2 cache*: Geometry and pixels processors share a 64 kB L2 read-only cache.
5. Power management unit, supporting power gating for all the other blocks.

Some devices in the Zynq UltraScale+ family also include a video codec unit in the form of specialized hardware block (i.e., as part of the FPGA resources), supporting simultaneous encoding/decoding of video and audio streams. Its combination with the Mali-400 GPU results in a very suitable platform for multimedia applications.

---

* OpenVG 1.0 is a royalty-free, cross-platform API for hardware accelerated two-dimensional vector and raster graphics.
† OpenGL ES is a royalty-free, cross-platform API for full-function 2D and 3D graphics on embedded systems.

All building blocks the processing system of Zynq UltraScale+ devices consists of are interconnected among themselves and with the FPGA fabric through AMBA AXI4 interfaces (described in Section 3.5.1.3).

Once hard and soft processors have been analyzed, it is important to emphasize that their features and performance (although extremely important*) are not the only ones to consider when addressing the design of FPSoCs. The resources available in the FPGA fabric (analyzed in Chapter 2) also play a fundamental role in this context.

Given the increasing complexity of FPSoC platforms, the availability of efficient software tools for design and verification tasks is also of paramount importance in the potential success of these platforms in the market. To realize how true this is, one has to just think about what it may take to debug a heterogeneous multicore FPSoC, where general-purpose and real-time OSs may have to interact (maybe also with some proprietary kernels) and share a whole bunch of hardware resources (memory and peripherals integrated in the processing system, implemented in the FPGA fabric, available there as specialized hardware blocks, or even implemented in external devices). Tools and methodologies for FPGA-based design are analyzed in Chapter 6, where special attention is paid to SoPC design tools (Section 6.3).

3.4 Other "Configurable" SoC Solutions

In previous sections, the most typical FPSoC solutions commercially available have been analyzed. They all have at least two common characteristics: the basic architecture, consisting of an FPGA and one or more embedded processors, and the fact that they target a wide range of application domains, that is, they are not focused on specific applications. This section analyzes other solutions with specific characteristics because either they do not follow the aforementioned basic architecture (some of them are not even based on FPGA and might have been excluded from this book, but are included to give readers a comprehensive view of configurable SoC architectures) or they target specific application domains.

3.4.1 Sensor Hubs

The integration in mobile devices (tablets, smartphones, wearables, and IoT) of multiple sensors enabling real-time context awareness (identification of user's context) has contributed to the success of these devices. This is due to

---

* A clear conclusion deriving from the analyses in Sections 3.2 and 3.3 is that the main reason for the fast evolution of FPSoC platforms in recent years is related to the continuous development of more and more sophisticated SMP and AMP platforms.

the many services that can be offered based on the knowledge of data such as user state (e.g., sitting, walking, sleeping, or running), location, environmental conditions, or the ability to respond to voice commands. In order for the corresponding apps to work properly, it is necessary to have in place an always-on context aware monitoring and decision-making process involving data acquisition, storage and analysis, as well as a high computational power, because the necessary processing algorithms are usually very complex.

At first sight, one may think these are tasks that can be easily performed by traditional microcontroller- or DSP-based systems. However, in the case of mobile devices, power consumption from batteries becomes a fundamental concern, which requires specific solutions to tackle it. Real-time management of sensors implies a high power consumption if traditional processing platforms are used. This gave rise to a new paradigm, sensor hubs, which is very rapidly developing. Sensor hubs are coprocessing systems aimed at relieving a host processor from sensor management tasks, resulting in faster, more efficient, and less power-consuming (in the range of tens of microwatts) processing. They include the necessary hardware to detect changes in user's context in real time. Only when the change of context requires host attention, it is notified and takes over the process.

QuickLogic specifically focuses on sensor hubs for mobile devices, offering two design platforms in this area, namely, EOS S3 Sensor Processing SoC (QuickLogic 2015) and Customer-Specific Standard Product (CSSP) (QuickLogic 2010).

EOS S3 is a sensor processing SoC platform intended to support a wide range of sensors in mobile devices, such as high-performance microphones, or environmental, inertial, or light sensors. Its basic architecture is shown in Figure 3.15. It consists of a multicore processor including a set of specialized hardware blocks and an FPGA fabric.

Control and processing tasks are executed in two processors, an ARM Cortex-M4F, including an FPU and up to 512 kB of SRAM memory, and a flexible fusion engine (FFE), which is a QuickLogic proprietary DSP-like (single-cycle MAC) VLIW processor. The ARM core is in charge of general-purpose processing tasks, and it hosts the OS, in case it is necessary to use one. The FFE processor is in charge of sensor data processing algorithms (such as voice triggering and recognition, motion-compensated heart rate monitoring, indoor navigation, pedestrian dead reckoning, or gesture detection). It supports in-system reconfiguration and includes a change detector targeting always-on context awareness applications.

A third processor, the Sensor Manager, is in charge of initializing, calibrating, and sampling front-end sensors (accelerometer, gyroscope, magnetometer, and pressure, ambient light, proximity, gesture, temperature, humidity, and heart rate sensors), as well as of data storage.

Data transfer among processors is carried out using multiple-packet FIFOS and DMA, whereas they connect with the sensors and the host processor mainly through SPI and I²C serial interfaces. Analog inputs connected to

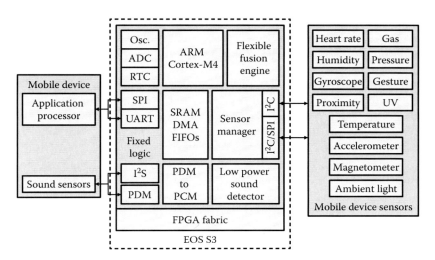

FIGURE 3.15
EOS S3 block diagram.

12-bit sigma-delta ADCs are available for battery monitoring or for connecting low-speed analog peripherals.

Given the importance of audio in mobile devices, EOS S3 includes resources supporting always-listening voice applications. These include interfaces for direct connection of integrated interchip sound (I²S) and pulse-density modulation (PDM) microphones, a hardware PDM to pulse-code modulation (PCM) converter (which converts the output of low-cost PDM microphones to PCM for high-accuracy on-chip voice recognition without the need for using CODECs), and a hardware accelerator based on Sensory's low power sound detector technology, in charge of detecting voice commands from low-level sound inputs. This block is capable of identifying if the sound coming from the microphone is actually voice, and only when this is the case, voice recognition tasks are carried out, providing significant energy savings.

Finally, the FPGA fabric allows the features of the FFE processor to be extended, the algorithms executed in either the ARM or the FFE processor to be accelerated, and user-defined functionalities to be added.

The CSSP platform was the predecessor of EOS S3 for the implementation of sensor hubs, but it can also support other applications related to connectivity and visualization in mobile devices. CSSP is not actually a family of devices, but a design approach, based on the use of configurable hardware platforms and a large portfolio of (mostly parameterizable) IP blocks, allowing the fast implementation of new products in the specific target application domains. The supporting hardware platforms are QuickLogic's PolarPro and ArcticLink device families.

PolarPro is a family of simple devices with a few specialized hardware blocks such as RAM, FIFO, and (in the most complex devices) SPI and

I²C interfaces. ArcticLink is a family of specific-purpose FPGAs that includes (in addition to the serial communication interfaces mentioned in Section 2.4.5) FFE and sensor manager processors, similar to those available in EOS S3 devices, and processing blocks to improve visualization or reduce consumption in the displays. The types and number of functional blocks available in each device depend on the specific target application. Figure 3.16 shows possible solutions for the three main application domains of CSSP: connectivity, visualization, and sensor hub:

- Connectivity applications are those intended to facilitate the connection of the host processor with both internal resources and external devices such as keyboards, headphone jacks, or even computers. FPGAs with hard serial communication interfaces (e.g., PolarPro 3E or ArcticLink) offer a suitable support to these applications.

- One of the most typical visualization problems in mobile devices is the lack of compatibility between display and main CPU bus interfaces. To ease interface interconnection, some devices from the ArcticLink family include specialized hardware blocks serving as bridges between the most widely used display bus interfaces (namely, MIPI, RGB, and LVDS). For instance, the ArcticLink III VX5 family includes devices with MIPI input and LVDS output, RGB input and LVDS output, MIPI input and RGB output, or RGB input and MIPI output.

- The hard blocks High Definition Visual Enhancement Engine (VEE HD+) and High Definition Display Power Optimizer (DPO HD+) in ArcticLink devices are oriented to improve image visualization and reduce battery power consumption. VEE HD+ allows dynamic range, contrast, and color saturation in images to be optimized, improving image perception under different lighting conditions. DPO HD+ uses statistical data provided by VEE HD+ to adjust brightness, achieving significant energy savings (it should be noted that in these systems, displays are responsible for 30%–60% of the overall consumption).

- CSSP supports sensor hub applications through ArcticLink 3 S2 devices, which include FFE and Sensor Manager processors (similar to those available in EOS S3 devices) and a SPI interface for connection to the host applications processor.

In addition to their specialized hardware blocks, there is a large portfolio of soft IP blocks available for the devices supporting the CSSP platform, called Proven System Blocks. These include data storage, network connection, image processing, or security-related blocks, among others. Finally, both EOS S3 and CSSP have drivers available to integrate the devices with different OSs, such as Android, Linux, and Windows Mobile.

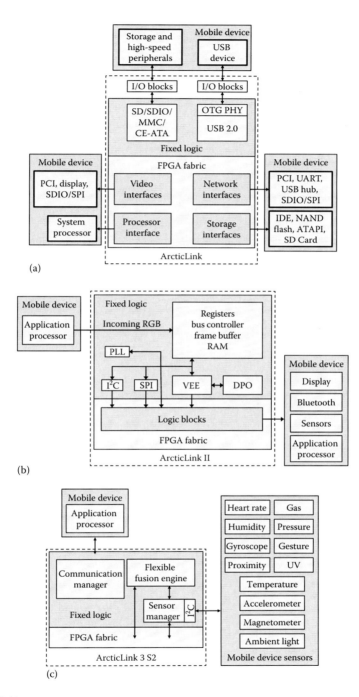

FIGURE 3.16
(a) Connectivity solution. (b) Visualization solution. (c) Sensor hub solution.

3.4.2 Customizable Processors

There are also non-FPGA-based configurable solutions offering designers a certain flexibility for the development of SoCs targeting specific applications. One such solution are customizable processors (Figure 3.17) (Cadence 2014; Synopsys 2015).

Customizable processors allow custom single- or multicore processors to be created from a basic core configuration and a given instruction set. Users can configure some of the resources of the processor to adapt its characteristics to the target application, as well as extend the instruction set by creating new instructions, for example, to accelerate critical functions.

Resource configuration includes the parameterization of some features of the core (instruction and data memory controllers, number of bits of internal buses, register structure, external communications interface, etc.), the possibility of adding or removing predefined components (such as multipliers, dividers, FPUs, DMA, GPIO, MAC units, interrupt controller, timers, or MMUs), or the possibility of adding new registers or user-defined components. This latter option is strongly linked to the ability to extend the instruction set, because most likely a new instruction will require some new hardware, and vice versa.

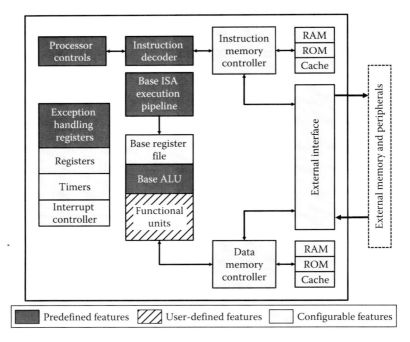

FIGURE 3.17
Customizable processors.

3.5 On-Chip Buses

One key factor to successfully develop embedded systems is to achieve an efficient communication between processors and their peripherals. Therefore, one of the major challenges of SoC technology is the design of the on-chip communication infrastructure, that is, the communication buses ensuring fast and secure exchange of information (either data or control signals), addressing issues such as speed, throughput, and latency. At the same time, it is very important (particularly when dealing with FPSoC platforms) that such functionality is available in the form of reusable IP cores, allowing both design costs and time to market to be reduced. Unfortunately, some IP core designers use different communication protocols and interfaces (even proprietary ones), complicating their integration and reuse, because of compatibility problems. In such cases, it is necessary to add glue logic to the designs. This creates problems related to degraded performance of the IP core and, in turn, of the whole SoC. To address these issues, over the years, some leading companies in the SoC market have proposed different on-chip bus architecture standards. The most popular ones are listed here:

- Advanced Microcontroller Bus Architecture (AMBA) from ARM (open standard)
- Avalon from Altera (open-source standard)
- CoreConnect from IBM (licensed, but available at no licensing or royalty cost for chip designers and core IP and tool developers)
- CoreFrame from PalmChip (licensed)
- Silicon Backplane from Sonics (licensed)
- STBus from STMicroelectronics (licensed)
- WishBone from OpenCores (open-source standard)

Most of these buses originated in association with certain processor architectures, for instance, AMBA (ARM processors), CoreConnect (PowerPC), or Avalon (Nios-II). Integration of a standard bus with its associated processor(s) is quite straightforward, resulting in modular systems with optimized and predictable behavior. Due to this, there is a trend, seen in the case of not only for chip vendors but also third-party IP companies, toward the use of technology-independent standard buses in library components, which ease design integration and verification.

In the FPGA market, AMBA has become the de facto dominating connectivity standard in industry for IP-based design because the leading vendors (Xilinx, Altera, Microsemi, QuickLogic) are clearly opting for embedding ARM processors (either Cortex-A or Cortex-M) within their chips. Other buses widely used in FPSoCs are Avalon and CoreConnect because of their association with the Nios-II and MicroBlaze soft processors, respectively. Wishbone is also used in some Lattice and OpenCores processors. These four buses are analyzed in detail in Sections 3.5.1 through 3.5.4.

3.5.1 AMBA

AMBA originated as the communication bus for ARM processor cores. It consists of a set of protocols included in five different specifications. The most widely used protocols in FPSoCs are Advanced eXtensible Interface (AXI3, AXI4, AXI4-Lite, AXI4-Stream) and Advanced High-performance Bus (AHB). Therefore, these are the ones analyzed in detail here, but at the end of the section, a table is included to provide a more general view of AMBA.

3.5.1.1 AHB

AMBA 2 specification, published in 1999, introduced AHB and Advanced Peripheral Bus (APB) protocols (ARM 1999). AMBA 2 uses by default a hierarchical bus architecture with at least one system (main, AHB) bus and secondary (peripheral, APB) buses connected to it through bridges. The performance and bandwidth of the system bus ensure the proper interconnection of high-performance, high clock frequency modules such as processors, on-chip memories, and DMA devices. Secondary buses are optimized to connect low-power or low-bandwidth peripherals, their complexity being, as a consequence, also low. Usually these peripherals use memory-mapped registers and are accessed under programmed control.

The structure of a SoC based on this specification is shown in Figure 3.18. The processor and high-bandwidth peripherals are interconnected through an AHB bus, whereas low-bandwidth peripherals are interconnected through an APB bus. The connection between these two buses is made through a bridge that translates AHB transfer commands into APB format and buffers all address, data, and control signals between both buses to accommodate their (usually different) operating frequencies. This structure allows the effect of slow modules in the communications of fast ones to be limited.

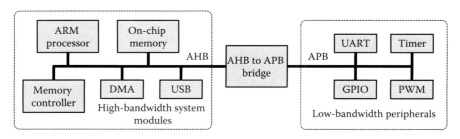

FIGURE 3.18
SoC based on AHB and APB protocols.

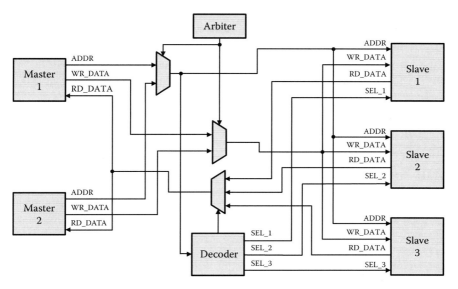

FIGURE 3.19
AHB bus structure according to AMBA 2 specification.

In order to fulfill the requirements of high-bandwidth modules, AHB supports pipelined operation, burst transfers, and split transactions, with a configurable data bus width up to 128 bits. As shown in Figure 3.19, it has a master–slave structure with arbiter, based on multiplexed interconnections and four basic blocks: AHB master, AHB slave, AHB arbiter, and AHB decoder.

AHB masters are the only blocks that can launch a read or write operation, by generating the address to be accessed, the data to be transferred (in the case of write operations), and the required control signals. In an AHB bus, there may be more than one master (multimaster architecture), but only one of them can take over the bus at a time.

AHB slaves react to read or write requests and notify the master if the transfer was successfully completed, if there was an error in it, or if it could not be completed so that the master has to retry (e.g., in the case of split transactions).

The AHB arbiter is responsible to ensure only one AHB master takes over the bus (i.e., starts a data transfer) at a time. Therefore, it defines the bus access hierarchy, by means of a fixed arbitration protocol.

Finally, the AHB decoder is used for address decoding, generating the right slave selection signals. In an AHB bus, there is only one arbiter and one decoder.

Operation is as follows: All masters willing to start a transfer generate the corresponding address and control signals. The arbiter then decides which master signals are to be sent to all slaves through the corresponding MUXs, while the decoder selects the slave actually involved in the transfer through another MUX. In case there is an APB bus, it acts as a slave of the corresponding bridge, which provides a second level of decoding for the APB slaves.

In FPSoCs using AHB, the processor is a master; the DMA controller is usually a master too. On-chip memories, external memory controllers, and APB bridges are usually AHB slaves. Although any peripheral can be connected as an AHB slave, if there is an APB bus, slow peripherals would be connected to it.

3.5.1.2 Multilayer AHB

AMBA 3 specification (ARM 2004a), published in 2003, introduces the multilayer AHB interconnection scheme, based on an interconnection matrix that allows multiple parallel connections between masters and slaves to be established. This provides increased flexibility, higher bandwidth, the possibility of associating the same slave to several masters, and reduced complexity, because arbitration tasks are limited to the cases when several masters want to access the same slave at the same time.

The simplest multilayer AHB structure is shown in Figure 3.20, where each master has its own AHB layer (i.e., there is only one master per layer). The decoder associated with each layer determines the slave involved in the transfer. If two masters request access to the same slave at the same time, the arbiter associated with the slave decides which master has the higher priority. The input stages of the interconnection matrix (one per layer) store the addresses and control signals corresponding to the pending transfers so that can be carried out later.

The number of input and output ports of the interconnection matrix can be adapted to the requirements of different applications. In this way, it is possible to build structures more complex than the one in Figure 3.20. For instance,

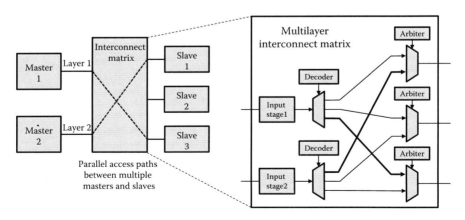

FIGURE 3.20
Multilayer interconnect topology.

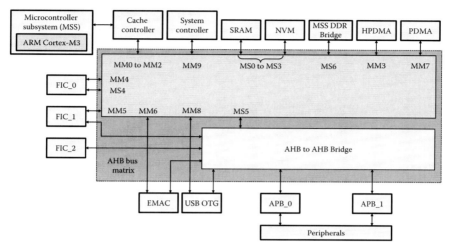

FIGURE 3.21
ARM Cortex-M3 core and peripherals in SmartFusion2 devices.

it is possible to have several masters in the same layer, define local slaves (connected to just one layer), or group a set of slaves so that the interconnection matrix treats them as a single one. This is useful, for instance, to combine low-bandwidth slaves.

An example of FPSoC that uses AHB/APB buses is the Microsemi SmartFusion2 SoC family (Microsemi 2013). As shown in Figure 3.21, it includes an ARM Cortex-M3 core and a set of peripherals organized in 10 masters (MM), 7 direct slaves (MS), and a large number of secondary slaves, connected through an AHB to AHB bridge and two APB bridges (APB_0 and APB_1). The AHB bus matrix is multilayer.

3.5.1.3 AXI

ARM introduced in AMBA 3 specification a new architecture, namely, AXI or, more precisely, AXI3 (ARM 2004b). The architecture was provided with additional functionalities in AMBA 4, resulting in AXI4 (ARM 2011). AXI provides a very efficient solution for communicating with high-frequency peripherals, as well as for multifrequency systems (i.e., systems with multiple clock domains).

AXI is currently a de facto standard for on-chip busing. A proof of its success is that some 35 leading companies (including OEM, EDA, and chip designers—FPGA vendors among them) cooperate in its development. As a result, AXI provides a communication interface and architecture suitable for SoC implementation in either ASICs or FPGAs.

AMBA 3 and AMBA 4 define four different versions of the protocol, namely, AXI3, AXI4, AXI4-Lite, and AXI4-Stream. Both AXI3 and AXI4 are

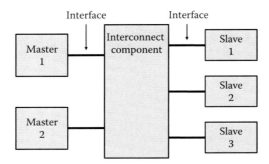

FIGURE 3.22
Architecture of the AXI protocol.

very robust, high-performance, memory-mapped solutions.* AXI4-Lite is a very reduced version of AXI4, intended to support access to control registers and low-performance peripherals. AXI4-Stream is intended to support high-speed streaming applications, where data access does not require addressing.

As shown in Figure 3.22, AXI architecture is conceptually similar to that of AHB in that both use master–slave configurations, where data transfers are launched by masters and there are interconnect components to connect masters to slaves.

The main difference is that AXI uses a point-to-point channel architecture, where address and control signals, read data, and write data use independent channels. This allows simultaneous, bidirectional data transfers between a master and a slave to be carried out, using handshake signals. A direct implication of this feature is that it eases the implementation of low-cost DMA systems.

AXI defines a single connection interface either to connect a master or a slave to the interconnect component or to directly connect a master to a slave. This interface has five different channels: read address channel, read data channel, write address channel, write data channel, and write response channel. Figure 3.23 shows read and write transactions in AXI.

Address and control information is sent through either the read or the write address channels. In read operations, the slave sends the master both data and a read response through the read data channel. The read response notifies the master that the read operation has been completed. The protocol includes an overlapping read burst feature, so the master may send a new read address before the slave has completed the current transaction. In this way, the slave can start preparing data for the new transaction while completing the current one, thus speeding up the read process. In write operations, the master sends data through the write data channel, and the slave

---

* Memory-mapped protocols refer to those where each data transfer accesses a certain address within a memory space (map).

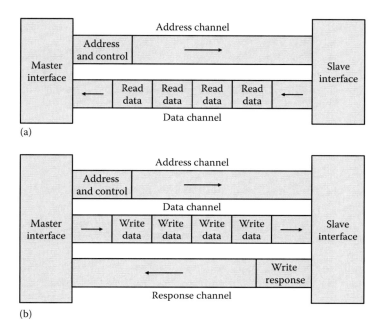

FIGURE 3.23
Read (a) and write (b) transactions in AXI protocol.

replies with a completion signal through the write response channel. Write data are buffered, so the master can start a new transfer before the slave notifies the completion of the current one. Read and write data bus widths are configurable from 8 to 1024 bits. All data transfers in AXI (except AXI4-Lite) are based on variable-length bursts, up to 16 transfers in AXI3 and up to 256 in AXI4. Only the starting address of the burst needs to be provided to start the transfer.

The interconnect component in Figure 3.22 is more versatile than the interconnection matrix in AHB. It is a component with more than one AMBA interface, in charge of connecting one or more masters to one or more slaves. In addition, it allows a set of masters or slaves to be grouped together, so they are seen as a single master or slave.

In order to adapt the balance between performance and complexity to different application requirements, the interconnect component can be configured in several modes. The most usual ones are shared address and data buses, shared address buses and multiple data buses, and multilayer, with multiple address and data buses. For instance, in systems requiring much higher bandwidth for data than for addresses, it is possible to share the address bus among different interfaces while having an independent data bus for each interface. In this way, data can be transferred in parallel at the same time as address channels are simplified.

Other interesting features of AXI are as follows:

- It supports pipeline stages (register slices in ARM's terminology) in all channels, so different throughput/latency trade-offs can be achieved depending on the number of stages. This is feasible because all channels are independent of each other and send information in only one direction.
- Each master–slave pair can operate at a different frequency, thus simplifying the implementation of multifrequency systems.
- It supports out-of-order transaction completion. For instance, if a master starts a transaction with a slow peripheral and later another one with a fast peripheral, it does not need to wait for the former to be completed before attending the latter (unless completing the transactions in a given order is a requirement of the application). In this way, the negative influence of dead times caused by slow peripherals is reduced. Complex peripherals can also take advantage of this feature to send their data out of order (some complex peripherals may generate different data with different latencies). Out-of-order transactions are supported in AXI by ID tags. The master assigns the same ID tag to all transactions that need to be completed on order and different ID tags to those not requiring a given order of completion.

We are just intending here to highlight some of the most significant features of AXI, but it is really a complex protocol because of its versatility and high degree of configurability. It includes many other features, such as unaligned data transfers, data upsizing and downsizing, different burst types, system cache, privileged and secure accesses, semaphore-type operations to enable exclusive accesses, and error support.

Today, the vast majority of FPSoCs use this type of interface, and vendors include a large variety of IP blocks based on it, which can be easily connected to create highly modular systems. In most cases, when including AXI-based IPs in a design, the interconnect logic is automatically generated and the designer usually just needs to define some configuration parameters.

The most important conclusion that can be extracted from the use of this solution is that it enables software developers to implement SoCs without the need for deep knowledge of FPGA technology, but mainly concentrating on programming tasks.

As an example, Xilinx adopted AXI as a communication interface for the IP cores in its FPGA families Spartan-6, Virtex-6, UltraScale, 7 series, and Zynq-7000 All Programmable SoC (Sundaramoorthy et al. 2010; Singh and Dao 2013; Xilinx 2015a). The portfolio of AXI-compliant IP cores includes a large number of peripherals widely used in SoC design, such as processors, timers, UARTs, memory controllers, Ethernet controllers, video controllers, and PCIe. In addition, a set of resources known as Infrastructure IP are also available to help in assembling the whole FPSoC. They provide features such as routing, transforming, and data checking.

Examples of such blocks are as follows:

- AXI Interconnect IP, to connect memory-mapped masters and slaves. It performs the tasks associated with the interconnect component by combining a set of IP cores (Figure 3.24):

    As commented earlier, AXI does not define the structure of the interconnect component, but it can be configured in multiple ways. The AXI Interconnect IP core supports the use models shown in Figure 3.25, which highlights the versatility and power of AXI for the implementation of FPSoCs.

- AXI Crossbar, to connect AXI memory-mapped peripherals.

- AXI Data Width Converter, to resize the datapath when master and slave use different data widths.

- AXI Clock Converter, to connect masters and slaves operating in different clock domains.

- AXI Protocol Converter, to connect an AXI3, AXI4, or AXI4-Lite master to a slave that uses a different protocol (e.g., AXI4 to AXI4-Lite or AXI4 to AXI3).

- AXI Data FIFO, to connect a master to a slave through FIFO buffers (it affects read and write channels).

- AXI Register Slice, to connect a master to a slave through a set of pipeline stages. In most cases, this is intended to reduce critical path delay.

- AXI Performance Monitors and Protocol Checkers, to test and debug AXI transactions.

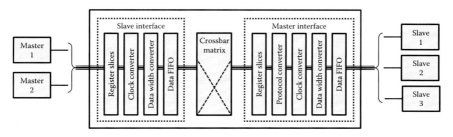

FIGURE 3.24
Block diagram of the Xilinx's AXI Interconnect IP core.

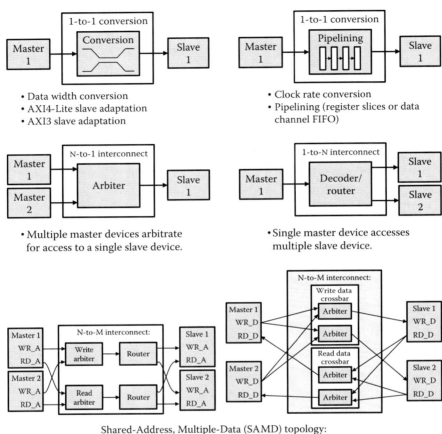

FIGURE 3.25
Xilinx's AXI Interconnect IP core use models.

In order for readers to have easy access to the most significant information regarding the different variations of AMBA, their main features are summarized in Table 3.2.

3.5.2 Avalon

Avalon is the solution provided by Altera to support FPSoC design based on the Nios-II soft processor. The original specification dates back to 2002, and a slightly modified version can be found in Altera (2003).

Avalon basically defines a master–slave structure with arbiter, which supports simultaneous data transfers among multiple master–slave pairs. When multiple

TABLE 3.2

Specifications and Protocols of the AMBA Communication Bus

Year	Spec.	Protocol	Aim and Features
1999	AMBA 2	AHB	Supports high-bandwidth system modules Main system bus in microcontroller usage Some features are • 32-bit address width and 8- to 128-bit data width • Single shared address bus and separate read and write data buses • Default hierarchical bus topology support • Supports multiple bus masters • Burst transfers • Split transactions • Pipelined operation (fixed pipeline between address/control and data phases) • Single-cycle bus master handover • Single-clock edge operation • Non-tri-state implementation • Single frequency system
		APB	Simple, low-power interface to support low-bandwidth peripherals Some features are • Local secondary bus encapsulated as a single AHB slave device • 32-bit address width and 32-bit data width • Simple interface • Latched address and control • Minimal gate count for peripherals • Burst transfers not supported • Unpipelined • All signal transitions are only related to the rising edge of the clock
		ASB	Obsolete

(Continued)

TABLE 3.2 (*Continued*)

Specifications and Protocols of the AMBA Communication Bus

Year	Spec.	Protocol	Aim and Features
2003	AMBA 3	AXI (AXI3)	Intended for high-performance memory-mapped requirements Key features: • 32-bit address width and 8- to 1024-bit data width • Five separate channels: read address, write address, read data, write data, and write response • Default bus matrix topology support • Simultaneous read and write transactions • Support for unaligned data transfers using byte strobes • Burst-based transactions with only start address issued • Fixed-burst mode for memory-mapped I/O peripherals • Ability to issue multiple outstanding addresses • Out-of-order transaction completion • Pipelined interconnect for high-speed operation • Register slices can be applied across any channel
		AHB-Lite	The main differences with regard to AHB are that it does not support multiple bus masters and extends data width up to 1024 bits
		APB	Includes two new features with regard to AMBA 2 specification, namely, wait states and error reporting
		ATB	*Advanced Trace Bus*: adds a data diagnostic interface to the AMBA specification for debugging purposes

(*Continued*)

TABLE 3.2 (Continued)

Specifications and Protocols of the AMBA Communication Bus

Year	Spec.	Protocol	Aim and Features
2011	AMBA 4	ACE	*AXI Coherency Extensions*: extends the AXI4 protocol and provides support for hardware-coherent caches. Enables correctness to be maintained when sharing data across caches
		ACE-Lite	Small subset of ACE signals
		AXI4	The main difference with regard to AXI3 is that it allows up to 256 beats of data per burst instead of just 16 It supports Quality of Service signaling
		AXI4-Lite	A subset of AXI4 intended for simple, low-throughput memory-mapped communications Key features: • Burst length of one for all transactions • 32- or 64-bit data bus • Exclusive accesses not supported
		AXI4-Stream	Intended for high-speed data streaming Designed for unidirectional data transfers from master to slave, greatly reducing routing Key features: • Supports single- and multiple data streams using the same set of shared wires • Supports multiple data widths within the same interconnect
		APB	Includes two new functionalities with regard to AMBA 3 specification, namely, transaction protection and sparse data transfer
2013	AMBA 5	CHI	*Coherent Hub Interface*: it defines the interconnection interface for fully coherent processors and dynamic memory controllers. Used in networks and serves

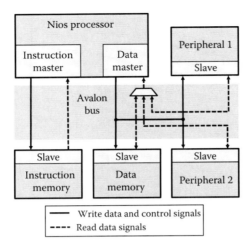

FIGURE 3.26
Sample FPSoC based on Altera's Avalon bus.

masters want to access the same slave, the arbitration logic defines the access priority and generates the control signals required to ensure all requested transactions are eventually completed. Figure 3.26 shows the block diagram of a sample FPSoC including a set of peripherals connected through an Avalon Bus Module.

The Avalon Bus Module includes all address, data, and control signals, as well as arbitration logic, required to connect the peripherals and build up the FPSoC. Its functionality includes address decoding for peripheral selection, wait-state generation to accommodate slow peripherals that cannot provide responses within a single clock cycle, identification and prioritization of interrupts generated by slave peripherals, or dynamic bus sizing to allow peripherals with different data widths to be connected. The original Avalon specification supports 8-, 16-, and 32-bit data.

Avalon uses separate ports for address, data, and control signals. In this way, the design of the peripherals is simplified, because there is no need for decoding each bus cycle to distinguish addresses from data or to disable outputs.

Although it is mainly oriented to memory-mapped connections, where each master–slave pair exchanges a single datum per bus transfer, the original Avalon specification also includes streaming peripherals and latency-aware peripherals modes (included in the Avalon Bus Module), oriented to support high-bandwidth peripherals. The first one eases transactions between streaming master and streaming slave to perform successive data transfers, which is particularly interesting for DMA transfers. The second one allows bandwidth usage to be optimized when accessing synchronous peripherals that require an initial latency to generate the first datum, but after that are capable of generating a new one each clock cycle (such as in the case of digital filters). In this mode, the master can execute a read request to the peripheral, then move to other tasks, and resume the read operation later.

As the demand for higher bandwidth and throughput was growing in many application domains, Avalon and the Nios-II architecture evolved to cope with it. The current Avalon specification (Altera 2015a) defines seven different interfaces:

1. Avalon Memory Mapped Interface (Avalon-MM), oriented to the connection of memory-mapped master–slave peripherals. It provides different operation modes supporting both simple peripherals requiring a fixed number of bus cycles to perform read or write transfers and much more complex ones, for example, with pipelining or burst capabilities. With regard to the original specification, maximum data width increases from 32 to 1024 bits.

   Like AMBA and many other memory-mapped buses, Avalon provides generic control and handshake signals to indicate the direction (read or write), start, end, successful completion, or error of each data transfer. Examples of such signals are "read," "write," or "response" in Figure 3.27. There are also specific signals required in advanced modes, such as arbitration signals in multimaster systems, wait signals to notify the master the slave cannot provide an immediate response to the request ("wait_request" in Figure 3.27), data valid signals (typical in pipelined peripherals to notify the master that there are valid data in the data bus, "read_data_valid" in Figure 3.27), or control signals for burst transfers.

2. Avalon Streaming Interface (Avalon-ST, Figure 3.28), oriented to peripherals performing high-bandwidth, low-latency, unidirectional point-to-point transfers. The simplest version supports single stream

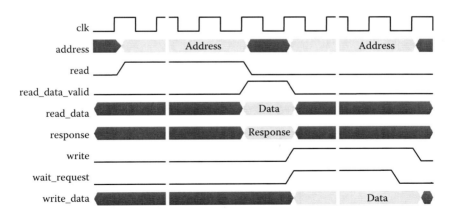

FIGURE 3.27
Typical read and write transfers of the Avalon-MM interface.

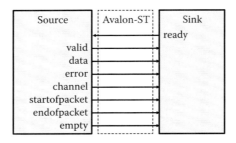

FIGURE 3.28
Avalon-ST interface signals.

of data, which only requires the signals "data" and "valid" to be used and, optionally, "channel" and "error." The sink interface samples data only if "valid" is active (i.e., there are valid data in "data"). The signal "channel" indicates the number of the channel, and "error" is a bit mask stating the error conditions considered in the data transfer (e.g., bit 0 and bit 1 may flag CRC and overflow errors, respectively).

Avalon-ST also allows interfaces supporting backpressure to be implemented. In this case, the source interface can only send data to the sink when this is ready to accept them (the signal "ready" is active). This is a usual technique to prevent data loss, for example, when the FIFO at the sink is full.

Finally, Avalon-ST supports burst and packet transfers. In packet-based transfers, "startofpacket" and "endofpacket" identify the first and last valid bus cycles of the packet. The signal "empty" identifies empty symbols in the packet, in the case of variable-length packets.

3. Avalon Conduit Interface, which allows data transfer signals (input, output, or bidirectional) to be created when they do not fit in any other types of Avalon interface. These are mainly used to design interfaces with external (off-chip) devices. Several conduits can be connected if they use the same type of signals, of the same width, and within the same clock domain.

4. Avalon Tri-State Conduit Interface (Avalon-TC), oriented to the design of controllers for external devices sharing resources such as address or data buses, or control signals in the terminals of the FPGA chip. Signal ·multiplexing is widely used to access multiple external devices minimizing the number of terminals required. In this case, the access to the shared terminals is based on tri-state signals. Avalon-TC includes all control and arbitration logic to identify multiplexed signals and give bus control to the right peripheral at any moment.

5. Avalon Interrupt Interface, which is in charge of managing interrupts generated by interrupt senders (slave peripherals) and notify them to the corresponding interrupt receivers (masters).

6. Avalon Reset Interface, which resets the internal logic of an interface or peripheral, forcing it to a user-defined safe state.

7. Avalon Clock Interface, which defines the clock signal(s) used by a peripheral. A peripheral may have clock input (clock sink), clock output (clock source), or both (for instance, in the case of PLLs). All other synchronous interfaces a peripheral may use (MM, ST, Conduit, TC, Interrupt, or Reset) are associated with a clock source acting as synchronization reference.

An FPSoC based on the Nios-II processor and Avalon may include multiple different interfaces or multiple instances of the same interface. Actually, a single component within the FPSoC may use any number and type of interfaces, as shown in Figure 3.29.

To ease the design and verification of Avalon-based FPSoCs, Altera provides the system integration tool Qsys (Altera 2015b), which automatically generates the suitable interconnect fabric (address/data bus connections, bus width matching logic, address decoder logic, arbitration logic) to connect a large number of IP cores available in its design libraries. Actually,

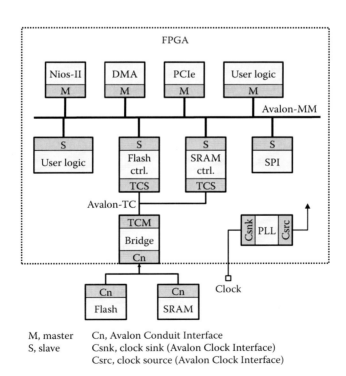

M, master   Cn, Avalon Conduit Interface
S, slave   Csnk, clock sink (Avalon Clock Interface)
              Csrc, clock source (Avalon Clock Interface)

FIGURE 3.29
Sample FPSoC using different Altera's Avalon interfaces.

Qsys also eases the design of systems using both Avalon and AXI and automatically generates bridges to connect components using different buses (Altera 2013).

3.5.3 CoreConnect

CoreConnect is an on-chip interconnection architecture proposed by IBM in the 1990s. Although the current strong trend to use ARM cores in the most current FPGA devices points to the supremacy of AMBA-based solutions, CoreConnect is briefly analyzed here because Xilinx uses it for the MicroBlaze (soft) and PowerPC (hard) embedded processors.

CoreConnect consists of three different buses, intended to accommodate memory-mapped or DMA peripherals of different performance levels (IBM 1999; Bergamaschi and Lee 2000):

1. Processor Local Bus (PLB), a system bus to serve the processor and connect high-bandwidth peripherals (such as on-chip memories or DMA controllers).
2. On-Chip Peripheral Bus (OPB), a secondary bus to connect low-bandwidth peripherals and reduce traffic in PLB.
3. Device Control Register (DCR), oriented to provide a channel to configure the control registers of the different peripherals from the processor and mainly used to initialize them.

The block diagram of the CoreConnect bus architecture is shown in Figure 3.30, where structural similarities with AMBA 2 (Figure 3.18) may be noticed. Same as AMBA 2, CoreConnect uses two buses, PLB and OPB, with different performance levels, interconnected through bridges.

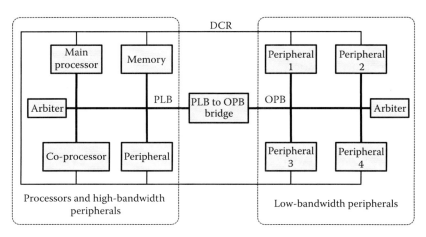

FIGURE 3.30
Sample FPSoC using CoreConnect bus architecture.

Both PLB and OPB use independent channels for addresses, read data, and write data. This enables simultaneous bidirectional transfers. They also support a multimaster structure with arbiter, where bus control is taken over by one master at a time.

PLB includes functionalities to improve transfer speed and safety, such as fixed- or variable-length burst transfers, line transfers, address pipelining (allowing a new read or write request to be overlapped with the one current being serviced), master-driven atomic operation, split transactions, or slave error reporting, among others.

PLB-to-OPB bridges allow PLB masters to access OPB peripherals, therefore acting as OPB masters and PLB slaves. Bridges support dynamic bus sizing (same as the buses themselves), line transfers, burst transfers, and DMA transfers to/from OPB masters.

Former Xilinx Virtex-II Pro and Virtex-4 families include embedded PowerPC 405 hard processors (Xilinx 2010a), whereas PowerPC 440 processors are included in Virtex-5 devices (Xilinx 2010b). In all cases, CoreConnect is used as communication interface. Specifically, PLB buses are used for data transfers and DCR for initializing the peripherals as well as for system verification purposes.

Although the most recent versions of the MicroBlaze soft processor (from 2013.1 on) use as main interconnection interfaces AMBA 4 (AXI4 and ACE) and Xilinx proprietary bus LMB, optionally, they can implement OPB.

3.5.4 WishBone

Wishbone Interconnection for Portable IP Cores (usually referred to just as Wishbone) is a communication interface developed by Silicore in 1999 and maintained since 2002 by OpenCores. Like the other interfaces described so far, Wishbone is based on a master–slave architecture, but, unlike them, it defines just one bus type, a high-speed bus. Systems requiring connections to both high-performance (i.e., high-speed, low-latency) and low-performance (i.e., low-speed, high-latency) peripherals may use two separate Wishbone interfaces without the need for using bridges.

The general Wishbone architecture is shown in Figure 3.31. It includes two basic blocks, namely, SYSCON (in charge of generating clock and reset signals) and INTERCON (the one containing the interconnections). It supports four different interconnection topologies, some of them with multimaster capabilities:

- Point to point, which connects a single master to a single slave.
- Data flow, used to implement pipelined systems. In this topology, each pipeline stage has a master interface and a slave interface.
- Shared bus, which connects two or more masters with one or more slaves, but only allows one single transaction to take place at a time.

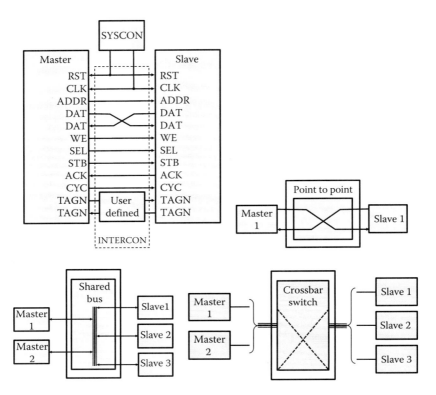

FIGURE 3.31
General architecture and connection topologies of Wishbone interfaces.

- Crossbar switch, which allows two or more masters to be simultaneously connected to two or more slaves; that is, it has several connection channels.

Shared bus and crossbar switch topologies require arbitration to define how and when each master accesses the slaves. However, arbiters are not defined in the Wishbone specification, so they have to be user defined.

According to Figure 3.31, Wishbone interfaces have independent address (ADR, 64-bit) and data (DAT, 8-/16-/32- or 64-bit) buses, as well as a set of handshake signals (selection [SEL], strobe [STB], acknowledge [ACK], error [ERR], retry [RTY], and cycle [CYC]) ensuring correct transmission of information and allowing data transfer rate to be adjusted for every bus cycle (all Wishbone bus cycles run at the speed of the slowest interface).

In addition to the signals defined in its specification, Wishbone supports user-defined ones in the form of "tags" (TAGN in Figure 3.31). These may be used for appending information to an address bus, a data bus, or a bus cycle. They are especially helpful to identify information such as data transfers, parity or error correction bits, interrupt vectors, or cache control operations.

Wishbone supports three basic data transfer modes:

1. Single read/write, used in single-data transfers.
2. Block read/write, used in burst transfers.
3. Read–modify–write, which allows data to be both read and written in a given memory location in the same bus cycle. During the first half of the cycle, a single read data transfer is performed, whereas a write data transfer is performed during the second half. The CYC_O signal (Figure 3.31) remains asserted during both halves of the cycle. This transfer mode is used in multiprocessor or multitask systems where different software processes share resources using semaphores to indicate whether a given resource is available or not at a given moment.

Wishbone is used in Lattice's LM8 and LM32, as well as in OpenCores' OpenRISC1200 soft processors, described in Sections 3.2.1 and 3.2.2, respectively.

References

Altera. 2002. Excalibur device overview data sheet. DS-EXCARM-2.0.
Altera. 2003. Avalon bus specification reference manual. MNL-AVABUSREF-1.2.
Altera. 2013. AMBA AXI and Altera Avalon Interoperation using Qsys. Available at: https://www.youtube.com/watch?v=LdD2B1x-5vo. Accessed November 20, 2016.
Altera. 2015a. Avalon interface specifications. MNLAVABUSREF 2015.03.04.
Altera. 2015b. Quartus prime standard edition handbook. QPS5V1 2015.05.04.
Altera. 2015c. Nios II classic processor reference guide. NII5V1 2015.04.02.
Altera. 2015d. Stratix 10 device overview data sheet. S10-OVERVIEW.
Altera. 2016a. Arria 10 hard processor system technical reference manual. Available at: https://www.altera.com/en_US/pdfs/literature/hb/arria-10/a10_5v4.pdf. Accessed November 20, 2016.
Altera. 2016b. Arria 10 device data sheet. A10-DATASHEET.
ARM. 1999. AMBA specification (rev 2.0) datasheet. IHI 0011A.
ARM. 2004a. Multilayer AHB overview datasheet. DVI 0045B.
ARM. 2004b. AMBA AXI protocol specification (v1.0) datasheet. IHI 0022B.
ARM. 2008. Cortex-M1 technical reference manual. DDI 0413D.
ARM. 2011. AMBA AXI and ACE protocol specification datasheet. IHI 0022D.
ARM. 2012. Cortex-A9 MPCore technical reference manual (rev. r4p1). ID091612.
Atmel. 2002. AT94K series field programmable system level integrated circuit data sheet. 1138F-FPSLI-06/02.
Bergamaschi, R.A. and Lee, W.R. 2000. Designing systems-on-chip using cores. In *Proceedings of the 37th Design Automation Conference (DAC 2000)*. June 5–9, Los Angeles, CA.
Cadence. 2014. Tensilica Xtensa 11 customizable processor datasheet.

IBM. 1999. The CoreConnect™ bus architecture.

Jeffers, J. and Reinders, J. 2015. *High Performance Parallelism Pearls. Multicore and Many-Core Programming Approaches*. Elsevier.

Kalray. 2014. MPPA ManyCore. Available at: http://www.kalrayinc.com/IMG/pdf/FLYER_MPPA_MANYCORE.pdf. Accessed November 20, 2016.

Kenny, R. and Watt, J. 2016. The breakthrough advantage for FPGAs with tri-gate technology. White Paper WP-01201-1.4. Available at: https://www.altera.com/content/dam/altera-www/global/en_US/pdfs/literature/wp/wp-01201-fpga-tri-gate-technology.pdf. Accessed November 23, 2016.

Kurisu, W. 2015. Addressing design challenges in heterogeneous multicore embedded systems. Mentor Graphics white paper TECH12350-w.

Lattice. 2008. Linux port to LatticeMico32 system reference guide.

Lattice. 2012. LatticeMico32 processor reference manual.

Lattice. 2014. LatticeMico8 processor reference manual.

Microsemi. 2013. SmartFusion2 microcontroller subsystem user guide.

Microsemi. 2016. SmartFusion2 system-on-chip FPGAs product brief. Available at: http://www.microsemi.com/products/fpga-soc/soc-fpga/smartfusion2#-documentation. Accessed November 20, 2016.

Moyer, B. 2013. *Real World Multicore Embedded Systems*: A Practical Approach. Elsevier–Newnes.

Nickolls, J. and Dally, W.J. 2010. The GPU computing era. *IEEE Micro*, 30:56–69.

NVIDIA. 2010. NVIDIA Tegra multi-processor architecture. Available at: http://www.nvidia.com/docs/io/90715/tegra_multiprocessor_architecture_white_paper_final_v1.1.pdf. Accessed November 20, 2016.

OpenCores. 2011. OpenRISC 1200 IP core specification (v0.11).

Pavlo, A. 2015. Emerging hardware trends in large-scale transaction processing. *IEEE Internet Computing*, 19:68–71.

QuickLogic. 2001. QL901M QuickMIPS data sheet.

QuickLogic. 2010. Customer specific standard product approach enables platform-based design. White paper (rev. F).

QuickLogic. 2015. QuickLogic EOS S3 sensor processing SoC platform brief. Datasheet.

Shalf, J., Bashor, J., Patterson, D., Asanovic, K., Yelick, K., Keutzer, K., and Mattson, T. 2009. The MANYCORE revolution: Will HPC LEAD or FOLLOW? Available at: http://cs.lbl.gov/news-media/news/2009/the-manycore-revolution-will-hpc-lead-or-follow/.

Sharma M. 2014. CoreSight SoC enabling efficient design of custom debug and trace subsystems for complex SoCs. Key steps to create a debug and trace solution for an ARM SoC. ARM White Paper. Available at: https://www.arm.com/files/pdf/building_debug_and_trace_multicore_soc.pdf. Accessed November 20, 2016.

Singh, V. and Dao, K. 2013. Maximize system performance using Xilinx based AXI4 interconnects. Xilinx white paper WP417.

Stallings, W. 2016. *Computer Organization and Architecture. Designing for Performance*, 10th edn. Pearson Education, UK.

Sundaramoorthy, N., Rao, N., and Hill, T. 2010. AXI4 interconnect paves the way to plug-and-play IP. Xilinx white paper WP379.

Synopsys. 2015. DesignWare ARC HS34 processor datasheet.

Tendler, J.M., Dodson, J.S., Fields Jr., J.S., Le, H., and Sinharoy, B. 2002. POWER4 system microarchitecture. *IBM Journal of Research and Development*, 46(1):5–25.

Triscend. 2000. Triscend E5 configurable system-on-chip family data sheet.
Triscend. 2001. Triscend A7 configurable system-on-chip platform data sheet.
Vadja, A. 2011. *Programming Many-Core Chips*. Springer Science + Business Media.
Walls, C. 2014. Selecting an operating system for embedded applications. Mentor Graphics white paper TECH112110-w.
Xilinx. 2008. Virtex-4 FPGA user guide UG070 (v2.6).
Xilinx. 2010a. PowerPC 405 processor block reference guide UG018 (v2.4).
Xilinx. 2010b. Embedded processor block in Virtex-5 FPGAs reference guide UG200 (v1.8).
Xilinx. 2011a. PicoBlaze 8-bit embedded microcontroller user guide UG129.
Xilinx. 2011b. Virtex-II Pro and Virtex-II Pro X platform FPGAs: Complete data sheet DS083 (v5.0).
Xilinx. 2014. Zynq-7000 all programmable SoC technical reference manual UG585 (v1.7).
Xilinx. 2015a. Vivado design suite—AXI reference guide UG1037.
Xilinx. 2015b. Xilinx collaborates with TSMC on 7nm for fourth consecutive generation of all programmable technology leadership and multi-node scaling advantage. Available at http://press.xilinx.com/2015-05-28-Xilinx-Collaborates-with-TSMC-on-7nm-for-Fourth-Consecutive-Generation-of-All-Programmable-Technology-Leadership-and-Multi-node-Scaling-Advantage. Accessed November 23, 2016.
Xilinx. 2016a. MicroBlaze processor reference guide UG984.
Xilinx. 2016b. Zynq UltraScale+ MPSoC overview data sheet DS891 (v1.1).

4

Advanced Signal Processing Resources in FPGAs

4.1 Introduction

Digital signal processing (DSP) is an area witnessing continuous significant advancements both in terms of software approaches and hardware platforms. Some of the most usual functionalities in this domain are digital filters, encoders, decoders, correlators, and mathematic transforms such as the fast Fourier transform (FFT). Most DSP functions and algorithms are quite complex and involve a large number of variables, coefficients, and stages. The key basic operation is usually provided by MAC units. Since high operating frequency and/or throughput are usually required, it is often necessary to use DSPs, whose hardware and instruction set are optimized for the execution of MAC operations or other features such as bit-reverse addressing (as discussed in Section 1.3.4.2). CPUs in DSPs are also designed to execute instructions in less clock cycles than in general-purpose processors. For many years, DSPs have been the only platforms capable of efficiently implementing DSP algorithms. However, in recent years, FPGAs have emerged as serious natural contenders in this area because of their intrinsic parallelism, their ability to very efficiently implement arithmetic operations, and the huge amount of logic resources available.

Since the advent of the first FPGAs in the 1980s, one of the main goals of vendors has been to ensure their devices are capable of efficiently implementing binary arithmetic operations (mainly addition, subtraction, and multiplication). This implies the need not only for specific logic resources but also for specialized interconnection resources, for example, for propagating carry signals or for chain connection of LBs, in order for propagation delays to be minimized and the number of bits of the operands to be parameterizable.

As FPGAs became increasingly popular, new application niches appeared requiring new specialized hardware resources. The availability of embedded memory blocks was particularly useful for the implementation of data acquisition and control circuits, avoiding (or at least mitigating) the need for external memories and reducing memory access times. After them, many other specialized hardware blocks were progressively included in each new family of devices, as described in detail in Chapter 2.

ALUs in conventional DSPs usually include from one to four MAC units operating in parallel. Their rigid architectures do not allow, for instance, the number of bits of the operands in a multiplication to be parameterized. Therefore, parallelism and bandwidth are inherently limited in these platforms, and increasing operating frequency is, in most cases, the only way of improving performance.

Let us consider as an example the implementation of an N-stage finite impulse response (FIR) filter in a DSP with four MAC units. From Figure 4.1a, it can be concluded that the algorithm has to be executed $N/4$ times for valid output data to be produced.

How can the same problem be solved using FPGAs? Thanks to the availability of abundant logic resources and the possibility of configuring them to operate in parallel, several approaches are feasible, from a fully series architecture (requiring N clock cycles to generate new output data) to a fully parallel one, like the one shown in Figure 4.1b, capable of generating new output data every clock cycle, or intermediate series-parallel solutions. This provides the designer with the flexibility to define different performance–complexity trade-offs by choosing a particular degree of parallelism. In addition, by using design techniques such as pipelining or retiming, extremely high-performance signal processing systems can be obtained.

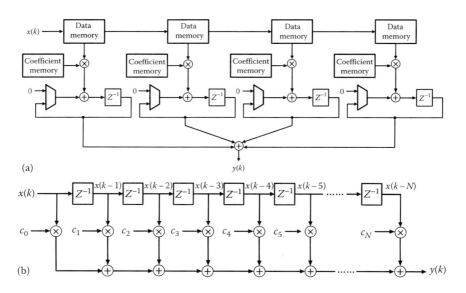

FIGURE 4.1
(a) FIR filter implemented with four MAC units and (b) fully parallel FIR filter.

Other advantages of the FPGA approach are the possibility to parameterize the size (number of bits per operand) of the arithmetic operators and the availability of different hardware structures to implement the MAC units.

The basic FPGA implementation of MAC units consists in building adders and multipliers using distributed logic, and combining them with embedded memory blocks, which act as accumulators and where coefficients are stored. However, in many cases, this solution implies the need for using many LBs, resulting not only in high resource consumption but also in long propagation delays, which limit operating frequency. Because of these issues, current FPGAs include specialized hardware blocks oriented to DSP applications, which are analyzed in the following sections. The simplest among these are hardware multipliers, but more complex ones (often referred to as DSP blocks) are also available.

4.2 Embedded Multipliers

The structure of a basic sample hardware 18-bit multiplier capable of performing both signed and unsigned operations is shown in Figure 4.2. In it, input and output registers allow (optionally) the operands and the result (36-bit, full-precision) to be memorized. In this way, for instance, the data and coefficients of a filter could be stored. Another advantage of these registers is for the straightforward implementation of efficient pipelining structures, taking advantage of the short delays associated with the dedicated connections inside the multiplier.

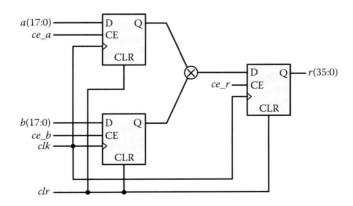

FIGURE 4.2
Embedded multiplier from Xilinx Spartan-3 devices. (From Xilinx, *Spartan-3 Generation FPGA User Guide: Extended Spartan-3A, Spartan-3E, and Spartan-3 FPGA Families: UG331 (v1.8)*, 2011.)

MAC units can be obtained by combining embedded multipliers, LBs (where additions may be implemented), and embedded memory (where input data and results are stored). This is the reason why embedded multipliers are usually placed adjacent to memory blocks, so routing among them is simplified, resulting in more efficient designs. Although 18 bits is not a usual data width in digital systems (it is not a power of 2), 18-bit multipliers are present in many FPGAs because they match a typical data width of memory blocks. Embedded memory data widths in FPGAs are usually multiples of 9, so information can be stored as sets of eight data bits plus one parity bit. However, if no data integrity checks are required, data can be stored in all nine bits. Therefore, it makes sense that arithmetic operators work with data widths that are multiples of 9.

Multipliers have associated resources for chain interconnection between adjacent blocks. In this way, it is possible to extend the number of bits of the operands or to build shift registers (which are usually required in DSP applications), by connecting input registers in a chain.

4.3 DSP Blocks

Considering signal processing over the years has been the most significant application of embedded multipliers, it is just natural that they evolved into more complex blocks, called DSP blocks, like the one in Figure 4.3, which includes all resources required to implement a MAC unit, eliminating the need for using distributed logic.

Different architectures exist for DSP blocks, but most of them share three main stages, namely pre-adder, multiplier, and ALU. Depending on the device, the ALU can just consist of an adder/subtractor or include

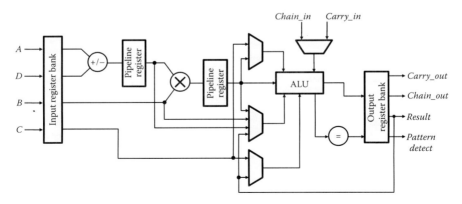

FIGURE 4.3
DSP block from Xilinx 7 Series.

additional resources (like in the case of Figure 4.3) aimed at giving the DSP block increased computation power (Altera 2011; Xilinx 2014; Lattice 2016; Microsemi 2016).

As in the case of multipliers, registers are placed at both the input and the output of the circuit in Figure 4.3, where interstage registers can also be identified. In this way, pipeline structures achieving very high operating frequencies can be implemented. In some DSP blocks from different FPGA families, double-registered inputs (consisting of two registers connected in a chain) are available, whereas other blocks include additional pipeline registers oriented to the implementation of systolic FIR filters. In this case, registers are placed at the input of the multiplier and at the output of the adder (which would be the input and output, respectively, of each stage of an FIR filter), to reduce interconnection delays. These registers are optional, so they can be bypassed if operation at the maximum achievable frequency is not required.

The significant amount of MUXes available provides the structure with many configuration possibilities. Thanks to them, it is possible to define different data paths and, in turn, different operating modes. In the case of Figure 4.3, the DSP block supports several independent functions, such as addition/subtraction, multiplication, MAC, multiplication and addition/subtraction, shifting, magnitude comparation, pattern detection, and counting. The selection inputs of the MUXes are usually accessible from distributed logic, allowing operating modes to be configured dynamically (normally in a synchronous way).

The pre-adder/subtractor may be used as an independent computing resource or to generate one of the input operands of the multiplier. This second alternative is useful for the implementation of some functionalities, for instance, the symmetric filter shown in Figure 4.4.

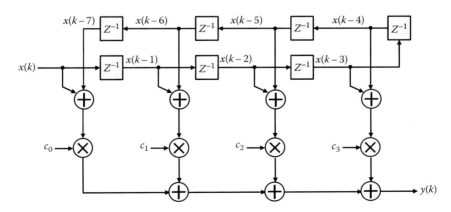

FIGURE 4.4
Eight-stage symmetric FIR filter.

The multiplier in Figure 4.3 operates with different input values (direct or registered inputs, data from the pre-adder or from the chain-connected adjacent block) and either stores the result in an intermediate register or sends it directly to the ALU, the "Result" output, or the "chainout" output.

Some DSP blocks in Altera FPGAs include a small "coefficient memory"* connected to one of the inputs of the multiplier, aimed at the optimized implementation of digital filters. Addressing is controlled from distributed logic, allowing filter reconfiguration to be dynamically performed during system operation. Thanks to this memory, there is no need for using embedded or distributed FPGA memory to store coefficients, therefore optimizing resource usage and reducing the time required to access coefficient values.

Regarding the ALU, it can perform arithmetic operations (addition, subtraction, and accumulation), logic functions, and pattern detection. When the accumulator is not used in conjunction with the multiplier, it can operate as an up/down synchronous counter. In some DSP blocks, the ALU can be divided into smaller units connected in chain and operating in parallel. For instance, a 48-bit ALU might operate as two 24-bit units, four 12-bit units, and so on. This feature is useful for the implementation of SIMD algorithms, so it is usually referred to as SIMD mode, an example of which is shown in Figure 4.5.

The pattern detection circuitry checks whether or not there is coincidence between one specific input of the DSP block (C in Figure 4.3) and the output of the ALU. Depending on the configuration, it is possible to implement other functions, such as convergent rounding, masked bit-wise operations, terminal count detection or autoreset of the counter, and detection of overflow/underflow conditions in the accumulator.

It is also possible to perform some combined multiplication–addition/subtraction operations with input data, for example, $(A \cdot B) \pm C$ or $(A + D) \cdot B \pm C$.

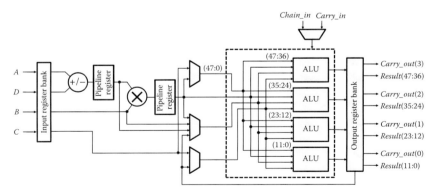

FIGURE 4.5
SIMD operating mode.

---

* "Internal Coefficient" in Arria 10 devices, which can store up to eight coefficients.

This allows, for instance, the result of the multiplication to be symmetrically rounded off to zero or to infinity.

As it may be expected, DSP blocks have dedicated lines for chain connection between adjacent blocks. In this way, the number of bits of the operands in arithmetic operations can be extended, and complex arithmetic functions or processing algorithms requiring multiple stages operating in parallel (e.g., digital filters) can be implemented. Same as embedded multipliers, DSP blocks are usually placed adjacent to embedded memory blocks.

The amount of DSP blocks available in a given FPGA depends on the target application profile. In devices oriented to signal processing, there may be some thousands of them,* achieving performances in the order of hundreds of GMAC/s. These very high computing speeds allow time multiplexing methods to be applied, in order for multiple operations of lower frequency to be carried out in a single DSP block. This results in semiparallel structures achieving very efficient trade-offs between resource usage and power consumption.

4.4 Floating-Point Hardware Operators

The embedded multipliers and DSP blocks described so far lack the ability to perform floating-point operations. Actually, most of these kinds of resources available in FPGAs are designed to operate in fixed point. This is a significant limitation in many cases because it implies the need for many signal processing designs to be adapted to work in fixed point, not only adding design burden but also creating potential accuracy problems. These issues are being overcome by the availability of new DSP blocks supporting the IEEE 754 floating-point standard.

The *IEEE Standard for Floating-Point Arithmetic* (IEEE 754) is the most widely used standard in floating-point computation circuits (IEEE 2008). It defines data formats, operations, and exceptions (such as division by zero, asymptotic functions, overflow, or inputs/outputs producing undefined or unrepresentable numbers, called NaN—Not a Number). The two basic data formats in IEEE 754 are simple (32-bit) and double (64-bit) precision. Any IEEE 754–compliant computing system must at least support simple precision operations.

The simple precision format consists of a *sign* bit (the most significant one in any data word), followed by 8 bits for the *exponent* (represented in *excess to* $2^{n-1} - 1$ format) and 23 bits for the *mantissa*, which is normalized so that it always starts with a nonzero bit. Therefore, in order for some operations (e.g., addition and subtraction) to be performed, it is first necessary to *align*

---

* Up to 3600 in Xilinx 7 Series devices.

the mantissas of the operands (so that the decimal separator is in the same position in all of them), then operate, and finally round off and normalize again the result. Actually, IEEE 754 specifies that alignment and normalization operations be done for each operation.

If fixed-point multipliers or DSP blocks are to be used for IEEE 754–compliant floating-point operations, alignment and normalization should be necessarily done using distributed logic in the FPGA fabric. This usually implies the need for barrel shifters of up to 48 bits (when working in single precision), which requires a large amount of logic and interconnection resources to be used, in turn negatively affecting operating frequency, to the extent that it may become the limiting factor in the performance of the whole processing system. Performance degradation is more significant as the complexity of the target algorithm grows, because of the need for executing alignment and normalization steps in all operations.

Currently, DSP blocks supporting IEEE 754–compliant single-precision operations are available in some FPGAs (Parker 2014; Sinha 2014; Altera 2016). As the sample block in Figure 4.6 shows, they include an adder and a multiplier, both IEEE 754 compliant, and some registers and MUXes that, like in the blocks described in Sections 4.2 and 4.3, are intended to allow high operating frequencies to be achieved and to provide configurability. Supported operating modes are addition/subtraction, multiplication, MAC, multiplication and addition/subtraction, vector one/two, and complex multiplication mode, among others.

In this case, alignment and normalization operations are carried out inside the DSP block itself, avoiding the need for using distributed logic resources with these purposes and, therefore, eliminating the aforementioned negative impact of these operations in performance. These blocks also include the logic resources required to detect and flag the exceptions defined by the IEEE 754 standard.

Figures 4.7 through 4.10 show some of the operating modes for floating-point arithmetic supported by the DSP block in Figure 4.6.

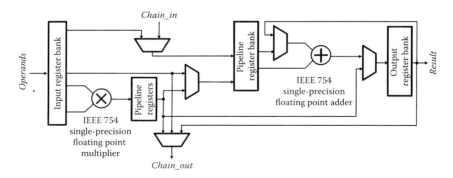

FIGURE 4.6
Variable Precision DSP Block from Altera Arria 10 FPGAs.

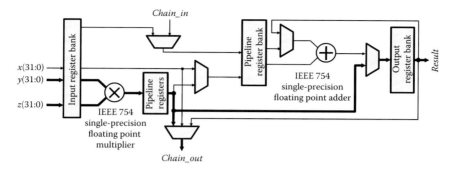

FIGURE 4.7
Multiplication mode: floating-point multiplication of input operands y and z.

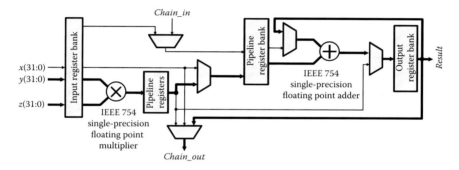

FIGURE 4.8
MAC mode: floating-point multiplication of input operands y and z, followed by floating-point addition/subtraction of the result and the previously accumulated value ($y \cdot z + acc$ or $y \cdot z - acc$).

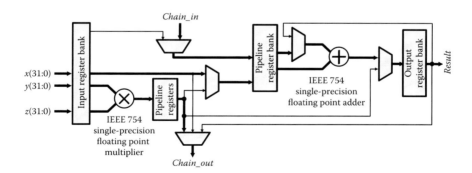

FIGURE 4.9
Vector two mode: simultaneous floating-point multiplication (whose result is sent to the following DSP block through the *chainout* output) and addition of the value received through the *chainin* input (from the previous DSP block) to the x input operand ($Result_n = x_n + chainin_n = x_n + chainout_{n-1} = x_n + y_{n-1} \cdot z_{n-1}$).

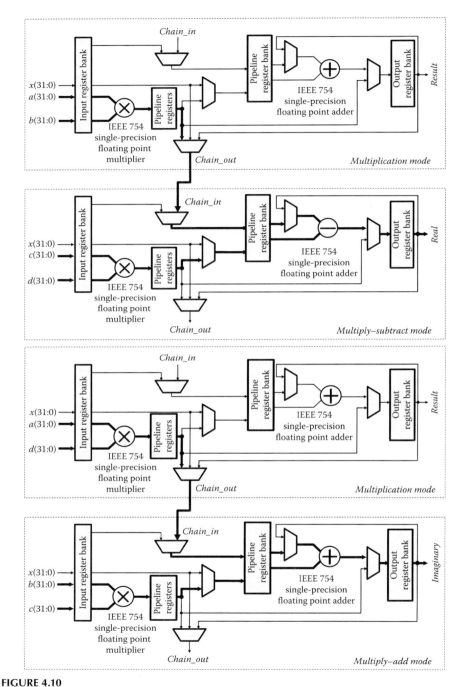

FIGURE 4.10
Complex multiplication mode: floating-point complex multiplication using four DSP blocks, according to the expression $(a + j \cdot b) \cdot (c + j \cdot d) = (a \cdot c - b \cdot d) + j \cdot (a \cdot d + b \cdot c)$.

Vendors provide sets of floating-point mathematic functions (many of which comply with specifications such as OpenCL 1.2) optimized for their implementation in these blocks.

In general, the design tools from the different vendors significantly automate the optimization and use of DSP resources available in their FPGAs. In this way, for applications without extremely demanding timing requirements, designers can easily develop fully functional systems without taking care of complex hardware issues, such as the internal topology of the blocks, pipeline acceleration, or time-division multiplexing techniques.

References

Altera. 2011. *Stratix IV Device Handbook (Vol. 1). DSP Blocks in Stratix IV Devices.* San Jose, CA.

Altera. 2016. *Arria 10 Core Fabric and General Purpose I/Os Handbook A10.* San Jose, CA.

IEEE. 2008. 754-2008—IEEE standard for floating-point arithmetic. Revision of ANSI/IEEE Std 754-1985.

Lattice. 2016. ECP5 and ECP5-5G family. Data Sheet DS1044 Version 1.6. Portland, OR.

Microsemi. 2016. *SmartFusion2 SoC and IGLOO2 FPGA Fabric: UG0445 User Guide.* Aliso Viejo, CA.

Parker, M. 2014. Understanding peak floating-point performance claims, Altera white paper WP-01222-1.0.

Sinha, U. 2014. Enabling impactful DSP designs on FPGAs with hardened floating-point implementation, Altera white paper WP-01227-1.0.

Xilinx. 2011. *Spartan-3 Generation FPGA User Guide: Extended Spartan-3A, Spartan-3E, and Spartan-3 FPGA Families: UG331 (v1.8).* San Jose, CA.

Xilinx. 2014. *7 Series DSP48E1 Slice. User Guide: UG479 (v1.8).* San Jose, CA.

5

Mixed-Signal FPGAs

5.1 Introduction

FPGAs were originally conceived as pure digital devices, not including any analog circuitry, such as input or output analog interfaces, which had to be built outside the FPGA whenever required. In contrast, analog circuitry is necessary for many FPGA applications (in general, but particularly in the case of industrial embedded systems) where, therefore, the need for ADCs and DACs is unavoidable. Even if control logic for external ADCs and DACs can be usually implemented using distributed logic inside the FPGA, eliminating the need for additional chips implementing glue logic and delays associated with external interconnections limit sampling or reconstruction frequency and may cause synchronization problems, thus having a negative impact on performance.

The solution to this drawback is obvious: to include ADCs and DACs inside FPGAs as hardware specialized blocks, like the ones discussed in Section 2.4. This not only mitigates bandwidth and synchronization problems, but also allows chip count to be reduced. As a consequence, some FPGA vendors decided to include analog front ends in some of their device families, giving rise to the so-called mixed-signal FPGAs (Xilinx 2011, 2014, 2015; Microsemi 2014a,b; Altera 2015). Also, some existing devices combine configurable analog and digital resources in the so-called field-programmable analog arrays (FPAAs) (Anadigm 2006).

The quite specific analog resources available in mixed-signal FPGAs are described throughout this chapter. Section 5.2 deals with ADC blocks, whereas analog sensors, and analog data acquisition and processing interfaces are described in Sections 5.3 and 5.4, respectively. Although FPAAs themselves are out of the scope of this book, hybrid FPGA–FPAA solutions are analyzed in Section 5.5.

5.2 ADC Blocks

Figure 5.1 shows the mixed-signal FPGA architectures of Altera MAX 10 and Xilinx 7 Series devices. They include up to two* 12-bit Successive Approximation Register (SAR) ADCs and share the following features:

- Maximum sampling rate: 1 Msps (minimum conversion time 1 μs).
- Single clock cycle conversion.
- Several multiplexed input channels (up to 18 in Altera devices with 2 ADCs and up to 17 in Xilinx ones). Channel 0 is a dedicated analog input, whereas all others are dual function (i.e., they can be configured as either analog inputs or general-purpose digital I/O pins).

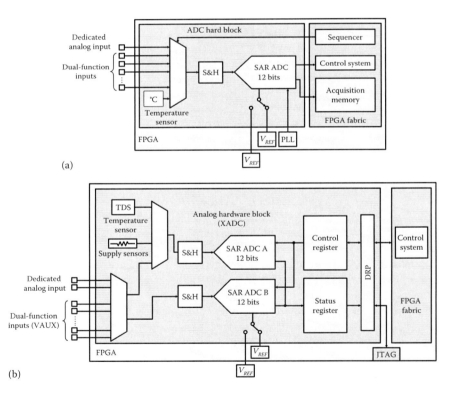

FIGURE 5.1
(a) ADC hard block from Altera's MAX 10 family and (b) XADC block from Xilinx 7 Series.

---

* All Xilinx 7 Series devices include two ADCs, whereas Altera's MAX devices may contain one or two, depending on the particular device.

- Support both single-ended and differential input signals.
- Support different operation modes, such as continuous sampling, conversion triggered by a specific event, and independent or simultaneous sampling (in devices with two ADCs).
- Support internal or external voltage references. Since ADC accuracy strongly depends on reference voltage, any ripple or noise affecting it negatively impacts conversion quality (e.g., conversion gain or signal-to-noise ratio). Since the internal voltage reference is usually the very supply voltage of the ADC (producing ratiometric conversion results), vendors recommend the use of external voltage references, for which it is easier to ensure better accuracy and lower temperature drift.
- Each ADC has an associated sample-and-hold (S&H) circuit or track-and-hold amplifier to ensure proper conversion. Some devices support configurable settling time.

On the other hand, there are differences in input voltage ranges and output configurations between the two families. In Altera MAX 10 devices, input voltages can be in the 0–2.5 or 0–3.3 V ranges, depending on the supply voltage, and the transfer function is unipolar (Figure 5.2). Input voltages are in the 0–1 V range in Xilinx 7 Series devices, whose transfer function can be unipolar or bipolar, in 2's complement (Figure 5.3).

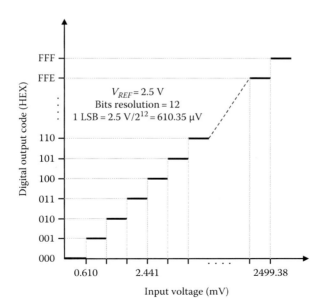

FIGURE 5.2
Unipolar ADC transfer function in Altera devices.

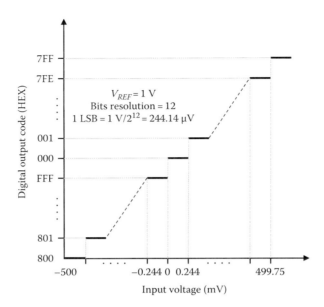

FIGURE 5.3
Bipolar ADC transfer functions in Xilinx 7 Series devices.

As can be clearly noticed in Figure 5.1, both hardware architectures are very similar. They consist of a set of input channels (coming from either external pins or internal signals), connected to the ADCs through multiplexing logic and S&H circuits, as well as of additional logic resources, aimed at configuring and controlling the ADCs and storing the results. For instance, control registers at the bottom of Figure 5.1 are used for storing the configuration parameters of the analog block, whereas status registers store converted data.

Control logic for hardware specialized analog blocks is implemented in the FPGA fabric. Specific communication resources are available for this purpose. For instance, the DRP (Dynamic Reconfiguration Port) in Figure 5.1b is a 16-bit synchronous read/write port, which enables access to control and status registers from the FPGA fabric. These registers are also accessible from the JTAG interface of the FPGA.

Specific IP cores are available for both MAX 10 and 7 Series devices to interact with the respective analog blocks. They are parameterizable soft controllers implementing several different predefined configurations and operating modes. They allow on-chip ADCs to be instantiated in a design, as well as clock signals and reference voltages to be configured, the input channel to be dynamically selected, maximum and minimum data values to be defined (and violation notifications to be generated if these are exceeded), and the whole data acquisition process to be managed in a way totally transparent to the designer, who does not need to care about low-level details.

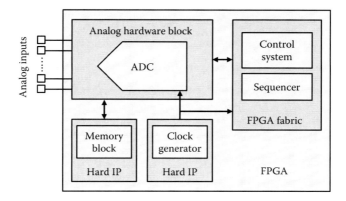

FIGURE 5.4
Common architecture of the control logic for analog resources in mixed-signal FPGAs.

Figure 5.4 shows the minimum set of resources required by the control logic:

- A control circuit (finite state machine), in charge of generating configuration and sampling signals (address, data, and transfer signals for configuration registers, start and end of conversion signals, handshake signals, etc.) so that they follow the required sequences and comply with the timing requirements specified by the vendor
- A sequencer to define the sampling sequence of input channels
- Memory resources to store data
- Clock and synchronization circuits

The control circuit and the sequencer are implemented in FPGA distributed logic, whereas acquisition memory may be implemented using either embedded or external memory blocks. Clock signals are generated using dedicated blocks (such as PLLs or DLLs) to ensure their stability and reduced skew.

These analog blocks support very diverse operating modes. In MAX 10 devices:

- It is possible to configure the order in which a set of input channels are sampled (i.e., an acquisition sequence). Acquisition sequences can be configured in single-trigger or continuous mode.
- In devices with two ADCs, each of them may be independently configured to have a different acquisition sequence, and acquisitions may be synchronous (using the same clock for both ADCs) or asynchronous (using different clock signals). In addition, simultaneous acquisition is supported in cases where the relative phase of input

signals must be kept unchanged. Simultaneous acquisition can only be implemented using the dedicated analog inputs (channel 0 of each ADC, as mentioned earlier), whose package routings are matched.

Both single-channel acquisition and acquisition sequences (automatic channel sequencer) are also possible in 7 Series devices, which also support single-trigger (single-pass mode) and continuous mode. Acquisitions in each ADC may be independent (one ADC acquires internal signals and the other external ones) or simultaneous. Differently from MAX 10 devices, in simultaneous acquisition mode, all external analog channels can be used. Since there are up to 16 dual-function pins (channels) available, up to eight simultaneous acquisition channels may be defined.

More specifically, in single-pass, automatic channel sequencer and continuous sequence modes, one ADC ("A") samples input signals (from temperature and voltage sensors) and the dedicated analog input, whereas the second ADC ("B") handles all other external channels. In simultaneous sampling mode, each ADC is connected to eight external signals, which are sampled in pairs (both ADCs operate in parallel), but it is also possible to include internal signals in the sampling sequence, associated with ADC "A" (in this mode, when sampling internal signals, ADC "B" is inactive). Finally, in independent ADC mode, ADC "A" samples internal signals, whereas ADC "B" samples the dedicated analog input and all other external channels.

Analog blocks in 7 Series devices also support external MUX mode (Figure 5.5), where (as the name indicates) an external input MUX (connected to the dedicated analog inputs) is used for channel multiplexing (channel selection logic is still generated by the embedded analog block). This is a useful option for designs where not enough I/O pins remain available once the digital logic has been defined. It has to be noted that the 17 input channels (differential inputs) supported by the analog block would actually consume 34 I/O pins.

Following the general philosophy behind configurable devices, analog resources in mixed-signal FPGAs are usually highly configurable. In addition to the aforementioned operating modes, other functionalities and parameters may be configured, even at run time. For instance, in 7 Series devices, each analog input can be independently configured to operate in unipolar or bipolar mode (to reduce common-mode noise), the output digital value may be obtained as the direct result of a single conversion or as the average of a set of samples (16, 64, or 256), and each ADC may be digitally calibrated to reduce gain and offset errors. Calibration is performed by connecting the ADC input to a known voltage, computing gain and offset errors, and generating the corresponding correction coefficients. Users can choose whether or not correction coefficients are applied by enabling or disabling the calibration option, respectively.

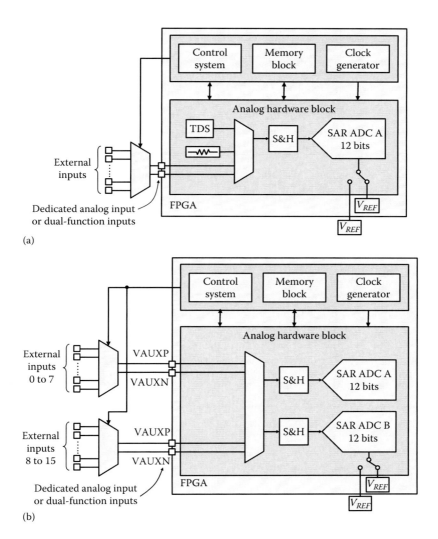

FIGURE 5.5
External (a) MUX and (b) simultaneous sampling modes.

5.3 Analog Sensors

In addition to ADCs, mixed-signal FPGAs usually include sensors that allow some of their operating parameters to be monitored. For instance, a sensor to measure die temperature is available in both Altera MAX 10 and Xilinx 7 Series devices. Monitoring die temperature allows to check if the device is working within an acceptable temperature range, hence helping to prevent damages due to excessive heating.

These sensors generate a voltage proportional to on-chip temperature, which, for instance, in the case of 7 Series devices is

$$V(T) = 10 \times \frac{kT}{q} \ln(10)$$

where
 k is the Boltzmann's constant ($1.38 \cdot 10^{-23}$ J/K)
 T is the temperature (K)
 q is the charge of the electron ($1.6 \cdot 10^{-19}$ C)

Voltage sensors are also available in 7 Series devices to measure on-chip power supply voltages (as shown in Figure 5.1). Both temperature and voltage sensors are connected to the input of the ADC through the input MUXes, and they are sampled in the same way as all other analog inputs. Usually, sampling of these signals is carried out by default, and alarms are generated even if the analog resources are not being used.

It is possible to define thresholds for these signals and generate alarms if these values are exceeded. For instance, some devices go to an inactive state when temperature or supply voltage go outside the acceptable ranges. Also, the speed of the fan cooling a device can be adjusted as a response to a temperature alarm.

In addition to the MAX 10 and 7 Series devices, FPGAs from other families also include less-performant analog resources. Altera Stratix V, Stratix IV, Arria V, and Arria V GZ FPGAs include a temperature sensor diode (TSD) with a built-in 8-bit (10-bit in Arria 10 devices) ADC circuitry to monitor die temperature. Xilinx Virtex-5, Virtex-6, UltraScale, UltraScale+, and Zynq UltraScale+ MPSoC families include an analog block, called System Monitor (SYSMON, with several versions available depending on the device), with embedded temperature and voltage sensors and the same number of external analog channels as in 7 Series devices, but with just a 10-bit, 20 ksps ADC.

5.4 Analog Data Acquisition and Processing Interfaces

Analog resources in Microsemi SmartFusion FPGAs (Microsemi 2014a) are more complex than those described so far. They build a subsystem for analog signal acquisition and processing, called Analog Compute Engine (ACE), which consists of three blocks (Figure 5.6): analog front-end interface, Sample Sequencing Engine (SSE), and Post-Processing Engine (PPE).

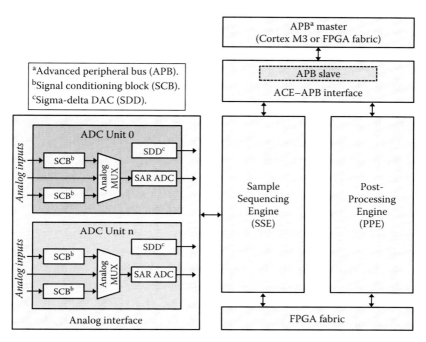

FIGURE 5.6
ACE analog subsystem from Microsemi SmartFusion family.

The analog front end includes signal conditioning circuits with S&H, analog MUXes, ADCs, and DACs. ADCs are SAR ones, with configurable resolution up to 12 bits (8, 10, or 12 bits). It supports simultaneous sampling of several ADCs. Reference voltage can be internal or external, in the 0–2.56 V range. To extend input voltage range, a prescaler is available with up to four different ranges.

Additional resources making a significant difference from other solutions are 24-bit delta–sigma DACs (as many as ADCs), current monitors based on differential-input, fixed-gain (50) amplifiers, temperature sensors, and high-speed analog comparators with configurable hysteresis thresholds. Single-ended analog inputs and outputs are multiplexed and demultiplexed, respectively. A very useful feature of these devices is that the embedded microcontrollers they include are equipped with dedicated interfaces to the analog circuitry, which in this regard acts as a slave peripheral of the microcontroller. Moreover, access and control of the analog resources can also be made from distributed logic without the need for using the microcontroller.

Given the complexity of the analog front end, a simple microcontroller (SSE) is available for configuring the parameters and operating modes of the different analog modules as well as for defining the sampling and conversion sequences of the input and output analog channels. Sampling sequences, resolution, and sampling times can be independently configured for each ADC. Simultaneous analog-to-digital conversion is supported as well as simultaneous updating of DAC outputs.

The PPE block is in charge of processing the signals from the ADCs. It uses FIFO memories to store data coming from each ADC and an ALU capable of performing calibration, threshold comparison, or other linear transforms. It can also be configured as a MAC unit, allowing low-pass filters to be implemented.

Thanks to the availability of SSE and PPE, there is no need to use embedded processors or distributed logic to perform the complex control and processing tasks associated with the analog part of the devices. Anyway, to facilitate high-level tasks, both SSE and PPE can generate interrupt requests to flag events related to calibration, the operation of the ADCs or the comparators, as well as general-purpose SSE events, or threshold comparison–related PPE events.

Another example of relatively complex mixed-signal FPGAs is the Microsemi Fusion family (Microsemi 2014b), whose architecture is shown in Figure 5.7. It includes up to 30 multiplexed analog inputs, a SAR ADC with configurable resolution (8, 10, or 12 bits) and sampling frequency up to 600 ksps, as well as temperature, voltage, and current sensors, and (as a

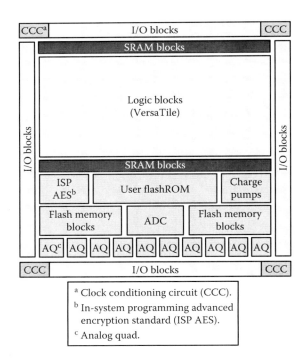

FIGURE 5.7
Architecture of Microsemi Fusion family.

significantly distinctive feature) up to 10 MOSFET gate driver outputs to control high-voltage external FETs. Same as other solutions, the reference voltage can be internal (2.56 V) or external (up to 3.3 V).

As shown in Figure 5.7, analog I/O resources are grouped in the so-called analog quad blocks, whose internal structure is shown in Figure 5.8. Each block includes three analog inputs (AV, AC, and AT) and a gate driver output pad (AG). They can be configured to operate in different modes, such as digital inputs, temperature or current monitor, or analog inputs with prescaler. Prescalers support different scaling factors to adapt to different ranges of positive (0–12 V) or negative (–12 to 0 V) input voltage.

Fusion devices include a TSD connected to channel 31 of the analog MUX, aimed at measuring internal chip temperature. In addition, the AT input of each analog quad can be connected to an external temperature sensor.

Current monitoring is carried out by connecting an external resistor of known value (typically less than 1 Ω) between two adjacent analog inputs

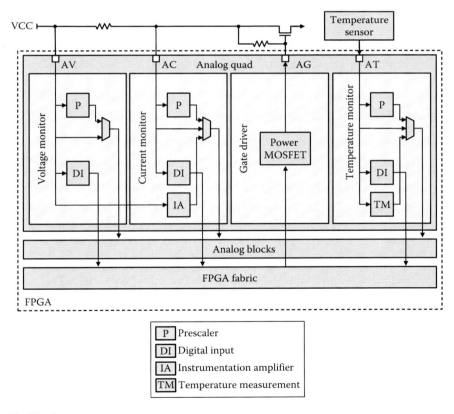

FIGURE 5.8
Block diagram of the analog quad blocks.

(AV and AC) and measuring the voltage drop between them. Operational amplifiers are available to amplify this voltage for improved current measurement accuracy.

5.5 Hybrid FPGA–FPAA Solutions

From previous sections, it is clear that, compared with logic resources, analog resources available in FPGAs are still quite limited. However, there is a trend for FPGA vendors to include in their most current devices an increasing number of analog blocks of increasing complexity.

For more than 20 years now, researchers and vendors have explored the feasibility of developing analog reconfigurable devices. Currently, commercial solutions already exist that combine configurable analog and digital resources. These are the so-called FPAAs, conceptually equivalent to FPGAs but oriented to analog applications. They consist of a set of analog blocks supporting a certain degree of configurability through the use of configurable analog blocks and digitally configurable interconnections to connect analog blocks among themselves and to I/O pins. Examples of such devices are Anadigm AN13x and AN23x families (Anadigm 2006).

Taking into account that the digital part of "pure" FPAAs is limited to interconnect and configuration resources, as well as to resources for the implementation of simple transfer functions, the detailed analysis of these devices is beyond the scope of this book. However, intermediate solutions between FPGAs and FPAAs exist. Such hybrid devices are available in Cypress PSoC 1, PSoC 3, PSoC 4, and PSoC 5LP family series (Cypress 2015). As the term PSoC suggests, these devices include embedded hardware processors, so they might have been analyzed in Chapter 3. However, considering that their most distinctive features are related to their analog part, we have decided to describe them here.

Figure 5.9 shows the architecture of the CY8C58LP family (PSoC 5LP series). It consists of three main blocks: Processor System, Digital System, and Analog System.

The Processor System includes, among other modules,

- A 32-bit ARM Cortex-M3 processor, capable of operating at up to 80 MHz (1.25 DMIPS/MHz)
- A Nested Vectored Interrupt Controller (NVIC) for fast interrupt handling, supporting up to 16 system exceptions and 32 interrupts
- Debug and trace modules accessible through JTAG or Serial Wire Debug interfaces

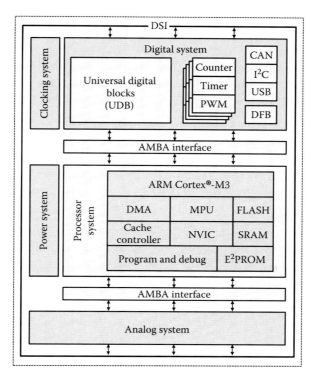

FIGURE 5.9
Architecture of the CY8C58LP family.

- Up to 256 kB of flash memory, 2 kB of EEPROM, and 64 kB of SRAM
- An external memory interface
- DMA and cache controllers

Connection of the Processor System with other parts of the device is made through a peripheral hub based on AMBA multilayer AHB interconnection scheme (described in Section 3.5.1.2).

The digital system consists of three main blocks:

1. An array of configurable logic blocks, called Universal Digital Blocks (UDBs)
2. Hard peripherals, including serial communication interfaces (CAN, USB, and I²C), timers, counters, and PWMs
3. A communication interface (digital system interface [DSI]) to interconnect reconfigurable logic, I/O pins, hard peripherals, interrupts, and DMA circuitry

Each UDB includes two PLDs (configurable structures much simpler than those in most current FPGAs, as introduced in Section 1.4), a datapath, and interconnection resources.

The datapath inside each UDB consists of an 8-bit single-cycle ALU and logic resources for comparison, shifting, and condition generation. It supports condition and signal propagation chains (e.g., carries) for the efficient implementation of arithmetic and shift operations. The datapath and the PLDs combine to build a UDB, and UDBs combine to build a UDB array.

Some devices in the CY8C58LP family also include a digital filter hardware block (DFB) as part of the digital system. The DFB includes a multiplier and an accumulator supporting 24-bit single-cycle MAC operations. To the best of authors' knowledge, no similar blocks exist in other devices to relieve the ARM Cortex-M3 core of this kind of highly bandwidth-consuming tasks.

Finally, the configurable analog system, which clearly separates these devices from any other current ones and whose structure is shown in Figure 5.10, consists of the following elements:

- A delta–sigma ADC whose default configuration is 16-bit resolution and 48 ksps, but is capable of also operating in other modes: 20, 12, or 8 bits and 187 sps, 192 ksps, or 384 ksps. It has a differential input, supports single and continuous sampling, and conversion start can be controlled either by software (by writing in a register) or hardware (through an external signal).

- Two 12-bit, 1 Msps SAR ADCs with single-ended or differential input.

- Four 8-bit DAC with voltage or current output. They support conversion rates up to 8 Msps for current output and 1 Msps for voltage output.

- Four analog comparators, whose outputs can be connected to four 2-input LUTs (allowing simple functions to be implemented) and, from them, to the digital system.

- Four programmable switched capacitor/continuous time (SC/CT) blocks, including one operational amplifier and a resistor network. With these elements, functionalities such as programmable gain amplifiers, transimpedance amplifiers, up/down mixers, S&H, and first-order analog to digital modulators, among others, may be built.

- Four general-purpose operational amplifiers supporting any voltage amplifier or follower configuration using either internal or external signals.

- A configurable interface for LCD displays, compatible with a wide variety of LCD displays.

- A capacitive touch sensing interface (CapSense subsystem in Figure 5.10) enabling capacitive measurements from devices such as proximity sensors, touch-sense buttons, and sliders.

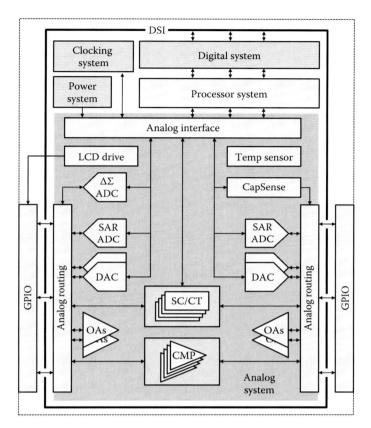

FIGURE 5.10
Analog system of CY8C58LP family.

- A temperature sensor to monitor internal device temperature.
- Internal high-precision reference voltages.
- Configurable resources to interconnect the different analog blocks as well as connect them with GPIOs. Interconnection resources are structured in global and local buses, MUXes, and switches.

For generation, synthesis, and distribution of clock signals, CY8C58LP devices include internal oscillators; specific (separate) clock frequency dividers for the digital, analog, and processor parts; and a fractional PLL with a working range from 24 to 80 MHz.

Cypress provides the PSoC Creator tool to support design of these devices. It eases configuration of both analog and digital interconnects, includes a library of predefined functions, and generates API interface libraries to set up communications between the process system and all other blocks in the device.

References

Altera. 2015. *MAX 10 Analog to Digital Converter User Guide: UG-M10ADC.*
Anadigm. 2006. AN13x series. AN23x series. *AnadigmApex dpASP Family User Manual.*
Cypress. 2015. PSoC 5LP: CY8C58LP family datasheet.
Microsemi. 2014a. *SmartFusion Programmable Analog User Guide.*
Microsemi. 2014b. *Fusion Family of Mixed Signal FPGAs (revision 6).*
Xilinx. 2011. *Virtex-5 FPGA System Monitor User Guide: UG192 (v1.7.1).*
Xilinx. 2014. *Virtex-6 FPGA System Monitor User Guide: UG370 (v1.2).*
Xilinx. 2015. *7 Series FPGAs and Zynq-7000 All Programmable SoC XADC Dual 12-Bit 1 MSPS Analog-to-Digital Converter User Guide: UG480 (v1.7).*

6

Tools and Methodologies for FPGA-Based Design

6.1 Introduction

Tools and methodologies for FPGA-based design have been continuously improving over the years in order for them to accommodate the new and extended functionality requirements imposed by increasingly demanding applications. Today's designs would take unacceptable extremely long times to be completed if tools coming from more than 20 years ago were used. The first important incremental step in accelerating design processes was the replacement of schematic-based design specifications by HDL descriptions (Riesgo et al. 1999).* On one hand, this allows complex circuits (described at different levels of abstraction) to be more efficiently simulated, and on the other hand, designs to be quite efficiently translated (by means of synthesis, mapping, placement, and routing tools) from HDLs into netlists, as a step previous to its translation into the bitstream with which the FPGA is configured (as described in Section 6.2.3.4).

Conventional synthesis tools were quite rapidly adopted by designers due to the productivity jump they enabled. At that point, it soon became apparent that FPGAs were very well suited to rapid prototyping and emulation flows because very little HDL code rework (or even none at all) was required in order to migrate designs initially implemented in FPGAs to other technologies. Either for prototyping or for final deployment, FPGAs rapidly increased their market share. As a consequence, and thanks to the improvement in manufacturing technologies, their complexity was continuously increased to cope with the ever-growing demand for more and more complex and integrated systems. This, in turn, contributed to higher market penetration, which pushed for additional complexity and expanded functionality, and so on.

---

* Most former techniques, particularly those based on schematic entry, are deprecated because of their low productivity. Therefore, they are not considered in this book.

The fast adoption of conventional synthesis tools as part of the natural design process for all types of digital hardware devices was not as fast, however, in the case of HLS tools (Cong et al. 2011). The difference between both types of tools resides in clock cycle explicitness. A conventional synthesizable HDL file mostly consists of descriptions where the transfers between memory elements can be directly and explicitly inferred from the code, clock cycle by clock cycle. In contrast, HLS tools start from descriptions that do not explicitly specify clock activity, but work at algorithmic level instead. The contribution or refinement HLS tools provide is their ability to allocate logic resources or operators and assign functions to such operators within the required time slots so that the algorithm may be mapped to a circuit with efficient resource sharing. Additionally, logic functions can be extended into optimized pipelined structures (so that the translation of such slots into clock cycles makes timing explicit), and clock speed can be optimized by adequately balancing critical paths within the pipelined structures. Regarding memories, different accessing schemes enable variable bandwidth adjustment so that it may adequately fit the functions being carried out by the logic reading/writing data from/to such memories. Finally, HLS tools also support two I/O types: memory mapped and stream based. These issues are analyzed in detail in Section 6.4.

Traditional or HLS tools alone cannot support the design of many of today's complex FPGA embedded systems. They need to be combined with platform-based tools that, in essence, automate different processes within a SoPC design flow (Sangiovanni-Vincentelli and Martin 2001). These tools combine standard components from integrated IP libraries with custom-made blocks designed using either conventional or HLS flows. Most current embedded systems are not fully customized designs, but rely on the combination of some standardized functions and interfaces with custom-made IP blocks. Therefore, module reuse and automated tools are mandatory in order to speed up the design process. Complex systems may be built with relatively little designer intervention if the design is based on library modules connected with standardized on-chip interfaces (described in Section 3.5). These tools provide, among many other features, module customization, automatic connection, automated memory map generation, as well as easy access to software code programmers by means of hardware abstraction layers for easy hardware/software interfacing. Users not familiar with this design methodology may be astonished to see how it allows highly complex designs to be readily obtained. For instance, a dual-core processor system with complex DMA schemes providing efficient access to a gigabit Ethernet media access control layer, plus some other I/O interfaces (such as SPI, I²C, USARTS, or GPIO), can be built in only a few hours.

Other tools are currently available whose design languages allow explicit parallelism to be described, aimed at achieving the maximum possible algorithm acceleration in HPC applications. They are based on OpenCL, which allows multithread parallelism to be mapped to heterogeneous computing platforms, such as FPGAs (Altera 2013; Xilinx 2014). In the last years, the main FPGA vendors are continuously releasing new specialized tools to ease the translation from OpenCL code into FPGA designs. These tools also provide ways for designs running in a host, usually a computer, to be accelerated by attaching one or more FPGA boards to it, often by means of PCIe connections (described in Section 2.4.4.1).

Increasingly, complex tools and design flows must necessarily be complemented with suitable validation and debugging methods. Verification can (and should) be done at early design stages, prior to circuit configuration, by means of simulation techniques. These techniques may be performed at functional level, to validate logic functionality, or after placement and routing, where accurate timing data are available to be annotated into the simulation. Very interestingly, as highlighted in Section 6.6.1.2, it is also possible to use integrated logic analyzers (embedded into the FPGA) for debugging purposes. These elements allow for combined hardware/software evaluation, which is very useful, especially for SoPC designs. Although some structured design validation techniques do exist, such as those derived from formal verification methods, they are not addressed in this book for two reasons: They are not specific to FPGA design and, to the best of authors' knowledge, there are no such commercial tools available for industrial use.

In the following sections, the different tools and design flows currently available are described in order of increasing complexity, which also corresponds with their evolution over time. Therefore, the conventional flow to transform netlists into bitstreams, based on the combination of register-transfer level (RTL) synthesis with back-end tools, is described in Section 6.2. Section 6.3 deals with the design flows and associated frameworks for SoPC systems, available for medium- and high-end FPGAs. HLS tools are discussed in Section 6.4. Contrary to what one might think, they appeared after SoPC platform-based designs, in part due to the slow adoption of these tools also in other areas, such as ASICs. In Section 6.5, tools for multithread acceleration in HPC applications are described. Finally, debugging tools and some second-order (or optional) tools available in many FPGA design frameworks are addressed in Section 6.6.

6.2 Basic Design Flow Based on RTL Synthesis and Implementation Tools

The combination of RTL synthesis and back-end tools is the core of the traditional synthesis-based FPGA design flow. These tools are essential in all other FPGA design flows since all of them eventually converge into this one. In essence, an RTL synthesizer takes as input HDL files with synthesizable code, which define the system to be designed. The output generated by the synthesizer is an intermediate representation of the circuit, where its basic structure can be identified but there is no link to any target technology. Back-end tools translate this generic structural representation into components available in the selected technology, map them into suitable locations within the FPGA fabric, and create the required interconnections by means of routing resources. If they succeed (which may not be the case, for instance, due to the lack of enough logic or interconnect resources in the target device), the configuration bitstream is generated. Finally, this may be used either to directly configure the FPGA or to program an external nonvolatile memory whose contents are loaded into the FPGA at power-up.

The main elements this flow consists of, as well as the main information coming in and out of the different tools, are depicted in Figure 6.1. Grayed elements represent the information to be provided by the user. Elements marked with an asterisk are optional.

The natural order to follow in this design flow starts with the creation of an HDL description of the system. This description is simulated in order to verify functional correctness. After functional simulation, RTL synthesis and back-end tools transform the HDL description into a placed and routed design. At this point, accurate timing information is available, enabling detailed timing simulations to be carried out. Finally, by creating the bitstream and configuring the FPGA with it, it is possible to verify the correct operation of the actual implementation. All these steps are discussed in detail in the following sections, which follow the aforementioned natural order.

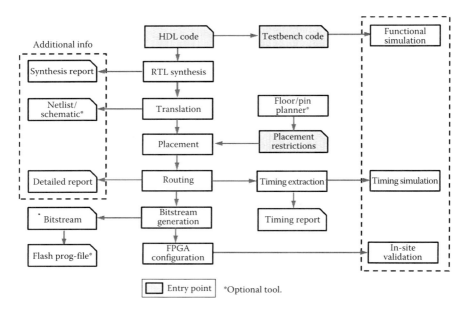

FIGURE 6.1
RTL synthesis and implementation tools' design flow.

6.2.1 Design Entry

The first stage of this flow corresponds to the entry of the required information into the design framework in order to specify the circuit to be designed. As shown in Figure 6.1, there are three entry points where external data have to be provided by the user, because they are design specific:

- The file(s) containing the HDL description(s) of the circuit to be designed for implementation in an FPGA.
- The file(s) describing the testbenches* for the device, under a somewhat realistic context. In many cases, device subsystems, if complex enough, should have their own testbenches, too. They are used for simulation purposes and are discussed in Section 6.2.2.
- The file defining the placement restrictions for I/O connections, which map signals in the design to I/O pins of the FPGA. Optionally, other placement restrictions and configuration attributes of internal components and signals within the design can be included. This file is used to guide placement and is therefore discussed in Section 6.2.3.3.

For medium- and high-complexity designs, HDL descriptions (entity/architecture pairs in the case of VHDL, modules in the case of Verilog) are better organized in a hierarchical way. Typically, the top-level descriptions just show the decomposition of the system into independent elements (modules), whose internal functionality is not described at this level, connected by signals. These descriptions simply place (instantiate) components and interconnect them, so they represent structure, not behavior. Ports within components are linked to signals by mappings associated with the instantiation. This approach is followed down the module hierarchy until functional descriptions are obtained for all circuit components, where their behavior can be identified. It is neither possible to simulate nor to synthesize a circuit until the behavior of all its components is described.

The RTL statements and description styles used to represent behavior are relatively simple but quite different from those of other languages, mainly because HDLs are not programming but description languages. They easily express concurrency (all hardware elements at architecture level are concurrent among them, so the order they are listed in the code is not significant) as well as data transfers in every active clock edge.

Concurrent statements, either conditional (e.g., *when...else*) or selective (e.g., *with...select*), represent combinational logic. They may define from simple Boolean expressions to more complex functional blocks, such as decoders, encoders, MUXs, and code translators. Data merging and splitting can be, respectively, described through data aggregations, for example, *"MacHeader*

---

* In VHDL notation. They are called textfixtures in Verilog. VHDL and Verilog are the two main existing HDLs.

<= *DestMAC & SourceMAC & TypeLength;"*, and vector subranges, for example, *"MSByte (7 downto 0) <= Word16 (15 downto 0);"*.

On the other side, synchronous sequential logic is typically described by means of processes where all synchronous signal assignments are conditionally executed within the clause *"if clk'event and clk='1' "* or similar. In this way, any such signal being assigned is equivalent to one flip-flop (single signal) or one register (vector signal). Processes run concurrently to other processes or other concurrent constructs, but they are "triggered" only when any of the signals in the so-called sensitivity list change. Once a process is triggered, statements within it are executed sequentially at "zero time," meaning that signal assignments do not take place in the sentence where they are defined, but all them occur together at the end of the process, with a minimum time delay ("delta"), which expresses causality (a process is triggered when signals on its sensitivity list change; signals assigned within the process change slightly after that). This behavior matches nicely with actual register updates taking place in the hardware at active clock edges.

Since signal assignments are not executed until the process is finished, all sentences within such clauses represent how the new values of the memory elements may be assigned according to the present values of signals and/or memory elements. This follows strictly the register-transfer level (from where RTL stands) rules these description types may have. Not all HDL clauses are acceptable for synthesis, but the RTL subset is.

If the conditions to assign new values to registers within a process are too complex to define, variables can be used. Opposite to signals, they change immediately as they are assigned in the sequential structure of a process or function. Variables are used, therefore, to approach algorithms, in the sense that register values are dependent on the combinations of variables and signals that are related algorithmically. As discussed in Section 6.4, HLS has a similar purpose. However, the difference is that the use of variables within a process does not modify the concept of register transfer, and all operations involving variables within a process that represents sequential synchronous logic take place within a clock cycle. Actually, complex algorithmic expressions may involve large critical paths by accumulating operators between register outputs and inputs. Contrary to this, HLS allows pipelining and other optimization techniques to be used since it does not start from a time-explicit description.

Design entry is, nowadays, automated and/or facilitated in many aspects to simplify designer's tasks. FPGA design frameworks are more integrated than ever, all options being available within the same environment. Editors for design entry may have features such as templates for basic constructs, syntax highlighting, automatic or aided indentation, on-the-fly syntax checking, code beautifiers, context search, and automatic block comment/uncomment. Also, some frameworks offer the possibility to have schematics automatically generated from structural descriptions, and navigation throughout the hierarchy of modules is enabled. For instance, module name-matching within working libraries allows automatic hierarchy identification to be achieved: All modules are sorted as a hierarchy tree with no need for configuration (i.e., no need to define methods for associating one component's module name with its description).

After design entry, a fearless designer might proceed straight into the synthesis and back-end process. However, it is normal that complete simulations of the important blocks, as well as for the whole design, are performed as an intermediate step. Simulation tools are described in Section 6.2.2.

6.2.2 Simulation Tools

Simulation is the preferred method to ensure that the description of a circuit matches its expected functionality. In simulations, a circuit must be set to work under the required conditions or, in other words, to receive a suitable and realistic set of input stimuli, allowing correct operation to be verified. The required stimuli sets are obtained through the generation of testbench descriptions. A testbench is an HDL file that contains an instantiation of the unit under test (UUT) and the elements that provide the stimuli to it. Optionally, testbench descriptions may include assertions to automatically check the fulfillment of some operating conditions.

Multivalued logic is available, and its use is strongly recommended. In this way, the ability to describe digital signals' behavior is extended from simply taking "0" or "1" strong logic values to many other situations that may occur in an actual circuit: "U" (unassigned), "X" (conflict), "Z" (high impedance), "H" (weak high), "L" (weak low), and "-" (don't care). This allows some problems in either the UUT or the testbench itself to be more easily identified. For instance, a not-initialized flip-flop in the design (because it does not have a reset signal) or in the testbench (because the reset signal is not asserted at the beginning) would produce unassigned values that would rapidly propagate through the design. This is due to the fact that simulators are conservative (or even pessimistic) in the sense that they intend to highlight any possible hazardous condition in the circuit, pointing designer's attention to them. Simulators can also highlight other common mistakes such as multiple assignments to signals coming from different sentences/processes, by setting the affected signals to "X."

For relatively simple circuits, stimuli are generated from processes that define the evolution of input signals over time. As simulation time elapses, these processes describe changes in input signals by using *wait for* (or similar) constructs, until all conditions are evaluated. Clocks are modeled in dedicated separate processes that take advantage of the possibility of reevaluating a process as soon as it finishes in order to achieve continuous operation. Testbench template file generation tools are capable of modeling this feature automatically. Other signals are typically grouped into different processes according to the origin of the incoming stimuli in order to mimic realistic operation. For instance, all signals involved in a communication channel are grouped into one process in order to produce input signal variations resembling the communication standard used in that channel (regarding issues such as timing, signal polarity, and coding). To this respect, the use of functions or procedures that carry out repetitive operations with different data is very helpful.

Processes, however, may not be the best approach for generating stimuli when UUTs are connected to (many) other elements, or there are intensive I/O operations, maybe requiring fine timing relationships among signals. In this case, the need arises for modeling the UUT and the other elements connected to it in such a way that stimuli for the different modules are provided at the right times. For instance, if an external memory is used, a model is needed for it, defining a more or less precise timing behavior (depending on the target timing precision for the particular simulation), in order for the interaction between both modules to be accurately described. Otherwise, the designer should have to precisely foresee when the UUT would issue a read transaction for the memory and generate the right data value at the right time according to the address the UUT is supposed to point to.

In general, this type of testbench modeling is required when there is strong module interaction or closed loops are present in the system. The need for such testbenches must be foreseen when accounting for design and validation efforts. Sometimes, it may be more difficult to model the environment of a circuit than the circuit itself. Simulation and verification times can never be neglected within the whole design process, but they are of particular significance for these heavily interacting systems.

Figure 6.2 shows sample block diagrams of the two aforementioned testbench types. Case (a) corresponds to a basic, process-based oriented testbench, where input stimuli come from a stand-alone process. Case (b) corresponds to a simulation with model components set together. The memory model interacts with the UUT by means of the corresponding bus signals. The communication module produces the necessary data sequences in its connections to the UUT, according to a process that generates data packets at the required moments.

As shown in Figure 6.1 and discussed in the following, simulations can be performed at two different stages of the design process, namely, at functional validation level and at timing verification level. In the first case,

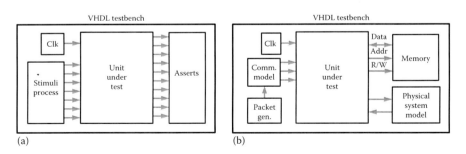

FIGURE 6.2
Testbench examples: (a) Stimuli process based; (b) side module based.

no timing information derived from the characteristics of the implementation is included, so it is often referred to as "ideal model" simulation. The second case corresponds to a stage where accurate timing information (at subclock cycle timing level) is available for analysis. Timing-accurate simulations can be performed, thanks to the data coming out from the timing analysis, which can be conducted after the place and route process (described in Section 6.2.3.3). The simulation model of the UUT is then fed back with data regarding delays in the internal logic and their associated connections inside the FPGA (including wires and switching blocks). Since this model contains a lot of information derived from the structure of the FPGA, simulation execution is much slower than in the case of the ideal model. This is the main reason why it is strongly advisable to perform an initial functional simulation of the circuit, to validate its functions at high level, as well as the overall activity and interactions, before proceeding to the place and route process and then timing simulation. Anyway, it is also possible to consider a "golden" reference testbench that can be applied to both models, equipped with asserts to automatically verify clauses ensuring that no deviations with respect to the target model occur during the synthesis and back-end design phases.

Not only very sophisticated simulation environments but also conventional ones within a classic design flow are nowadays equipped with feature-rich visualization and analysis tools, which significantly contribute to simplifying design validation. Data can be visualized as individual signal lines or grouped into buses, using different digital representations (such as binary, hexadecimal, or ASCII), or even as analog signals displayed in an oscilloscope-like format. Signals exhibiting nontypical or unexpected behaviors (e.g., taking an "X" value) are represented using different attention-calling colors. Signal navigation for long simulation runs can be accelerated by all kinds of zooming and panning. Navigation may be also done by selecting a signal and checking its evolution transition by transition; in this way, there is no need to look for specific values in a signal across long periods of time; the visualization tool can be asked to move forward (or backward) and find that small "cycle" almost hidden among all other signals. It is also possible to select the signals to be traced by navigating through the hierarchies of components and processes.

The simulation process can be enhanced with features that allow more realistic results to be achieved, execution to be accelerated, or interacting discrete- and continuous-time systems to be analyzed together. The resulting simulation approaches can be summarized in the following categories, addressed in subsequent sections:

- Interactive simulation
- Mixed-mode simulation
- HIL verification

6.2.2.1 Interactive Simulation

Simulations can be made interactive (and with customized graphical interfaces) for the sake of building virtual models that resemble as much as possible the appearance and the interactivity among elements, in particular when interaction with humans is to be verified/validated. Some simulators offer these possibilities by means of specific script languages (such as C or other standard programming languages), native tools embedded into other frameworks, or even communication sockets enabling distributed virtual or remote simulations. If fast-enough simulation platforms are available (e.g., combining powerful processors with simple and fast system models), interactivity can be made somehow similar to "real-life" behavior, for instance, allowing user interfaces to be validated and "mock-ups" of products to be made well before they are actually produced or built. However, the use of this technique is not advisable from the design validation viewpoint since unpredictable human interaction makes experiments lose repeatability.

6.2.2.2 Mixed-Mode Simulation

Mixed-mode simulation is a technique to be considered when the embedded system to be designed is part of a control loop, where the system to be controlled is to be modeled in continuous time rather than as a discrete-event system. In this case, it is possible to combine discrete-event simulators (such as HDL simulators for the required digital designs) with continuous-time simulators, which are effective for either analog circuits or any other physical systems modeled with continuous signals. For instance, when designing a motor controller, its behavior can be more realistically analyzed if the discrete events coming out of it are converted to analog signals and applied to a motor model so that both elements can be jointly simulated at the same time. There are also some specific HDLs targeting mixed-mode simulation, such as VHDL-AMS, that could be useful for such type of simulation.

6.2.2.3 HIL Verification

The evolution of the features and performance of simulation platforms currently enables the combination of emulation and simulation tasks through the use of HIL techniques. This approach is similar to mixed-mode simulation, but instead of a model of the physical system, the real system itself is used. A necessary condition for HIL platforms to provide realistic results is that the execution of the model being simulated has to be as fast as the real system it interacts with. This is not the case, of course, for systems with hard real-time requirements, but it is still feasible in a wide range of applications, and its use is being adopted in many design and verification flows.

Some companies include in their simulators features intended to support HIL operation, providing suitable interfaces between the simulation and emulation domains in a similar way as required in mixed-mode simulation for the interaction between the digital and the analog/physical models. However, since there is no standardization so far in this respect, the mixed emulated-simulated scenarios have to be customized on a case-by-case basis, usually implying a considerable amount of work. Although, as a consequence, the decision on whether to use this technique or not is strongly application dependent, it is becoming a very interesting possibility for many embedded systems in industrial applications, as discussed in Chapter 9.

6.2.3 RTL Synthesis and Back-End Tools

The validation of the functional description of a system in synthesizable HDL code is the green flag to proceed to the synthesis and implementation of the design. As discussed in Section 6.2 (and highlighted in Figure 6.1), the transformation from the HDL description of a circuit to the corresponding FPGA programming bitstream consists of several steps, namely, RTL synthesis (analyzed in detail in Section 6.2.3.1), translation into the target technology (Section 6.2.3.2), placement and routing (Section 6.2.3.3), and bitstream generation (Section 6.2.3.4). In addition to HDL descriptions, the specification of constraints for guided placement and some other parameters may be required to configure and guide the synthesis and implementation processes.

6.2.3.1 RTL Synthesis

This step is the most important of the basic design flow, where the logic elements that will actually perform the functions described in the HDL input file(s) are obtained. All elements described in the file(s) are translated into an intermediate representation, where operators that process signals are placed in between registers that memorize signal values from one clock cycle to the next. This characteristic is the one from which the name RTL (from register-transfer level) given to this type of synthesis is derived.

The operation of an RTL synthesizer can be explained from the state and output equations that define the evolution of any sequential system, which may be expressed as

$$Q_{t+1} = f(Q_t, X_t) \quad \text{(state equations)}$$

and

$$Y_t = g(Q_t, X_t) \quad \text{(output equations)}$$

where

Q are state variables

X are inputs

Y are outputs

Subindex "*t*" denotes current time

"*t+1*" denotes next time step (according to the discrete nature of time in synchronous sequential systems)

On the one side, an RTL synthesizer translates the expressions inside the clauses conditioned to work only at active clock edges (e.g., within *clk'event* clauses inside the processes that model sequential logic in VHDL) into state equations. On the other side, concurrent sentences are translated into output equations, that is, combinational functions that determine the value of the outputs at any time, given the state and the input values at the very same time.

Figure 6.3 illustrates with an example how a design (a simple binary counter) is gradually transformed from its description in the HDL input file to the final circuit implementation into the FPGA fabric. The VHDL process describing the behavior of the counter is shown in Figure 6.3a.

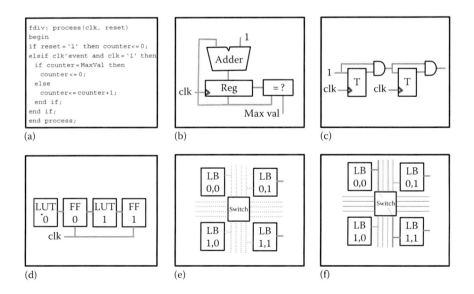

FIGURE 6.3

Stages during synthesis and implementation: (a) HDL code; (b) abstract model; (c) after Boolean optimization; (d) mapped into target technology; (e) placed design; (f) placed and routed design.

The assignments representing changes in the state of the counter that may take place at each active edge of the clock signal "clk" (i.e., conditioned to the occurrence of *"clk'event and clk = '1' "*) are

- *"counter <= '0';"*, which makes the counter roll over when the MaxVal value has been reached
- *"counter <= counter + 1;"*, which increases the counter value otherwise

Since the state of the counter is assigned with a synchronous operation, a register is required. Therefore, the translation of these sentences into the intermediate representation of the circuit shown in Figure 6.3b produces a register and a binary adder, which adds 1 to the current value of the register and transfers it into its next value if the counter is not at *MaxVal* value (a comparator is used to check this condition). Otherwise, the register synchronously rolls over to zero. This "synchronous clear" function must have higher priority than the "increment" function. This priority is explicitly expressed in the code, where the roll-over condition is evaluated in the *if* clause and the increment condition is evaluated in the corresponding *else* part.

At this point, a significant question arises. According to the specification, the HDL code has been transformed into a register, an adder, and a comparator. Does this mean that the hardware structure of a counter implemented from an HDL description is different from the well-known one in Figure 6.3c, typically taught in basic digital electronics courses? Apparently yes, but the real answer is no. After the translation into the intermediate representation, it is time for Boolean optimization. In this step, logic is as much simplified as possible, unnecessary logic is removed, redundant logic is reduced when identified as such (which is only possible to a certain extent in complex circuits), and, if specified, logic is arranged to fulfill timing constraints. These constraints are specified by the designer, and, in their simplest form, they just consist of a minimum operating frequency, which determines the maximum critical path delay of the design.

How is the structure in Figure 6.3c inferred from Figure 6.3b? As can be clearly seen, the adder has a constant integer value of 1 in one of its inputs, that is, a binary combination "00...0001." The 0s in this input greatly simplify the adder, the series AND gates that appear in the reduced circuit being a resemblance of the carry propagation chain. The transformation from D flip-flops (the usual ones for registers) to T flip-flops comes from the fact that the basic addition of two bits is equivalent to an XOR gate ($0 + 0 = 0$, $0 + 1 = 1$, $1 + 0 = 1$, $1 + 1 = 0$; "+" being, in this case, the arithmetic operator, not the logic one), and a T flip-flop is obtained from a D flip-flop by feeding back the output of the register by means of an XOR gate. Finally, if the counter's count cycle is a power of 2, the comparator disappears; otherwise, it would appear as an AND gate that will trigger register rollover.

6.2.3.2 Translation

The result of the synthesis is an intermediate generic representation with reduced logic expressions. However, this is not necessarily equivalent to having a minimized circuit built with conventional logic gates, nor are these gates size optimized for this target. Actually, the target technology might use building blocks that are far from being these logic gates, such as in the case of FPGAs, where LUTs and flip-flops are the basic constituents of the logic (as analyzed in Section 2.3.1). Therefore, in FPGA design, logic expressions must be, in principle, implemented with LUTs (specialized hardware blocks may also be used), in such a way that the same functionality is obtained, but using the elements that allow the circuit to be mapped into the target FPGA device.

With the advent of proprietary synthesizers for some FPGA families, the synthesis and translation steps are more tightly linked, since the way some structures are mapped into specialized resources, such as memories or DSP blocks, requires some knowledge of the underlying technology in order not to miss the possibility of using them. In some cases, the inference of some elements needs to follow a specific syntax for the synthesizer to recognize them. In other cases, the possibility exists to use attributes in the code (or synthesis and optimization options) to guide the tools to infer some specific elements and choose among different implementation possibilities. This is the case, for instance, for memories. The designer may, for instance, decide whether a memory in his/her design is to be mapped into embedded memory blocks or distributed along the logic (among other possibilities). Tools, in principle, should be able to choose a right solution (if it exists) so that the circuit fits into the target FPGA, but in case the system needs to be "fine-tuned," as for speed optimization or detailed positioning and mapping, the designer may decide to instantiate primitive technology–specific blocks, such as an LUT, to have full control on the mapping. This may be combined with placement restriction specifications, which are analyzed in Section 6.2.3.3.

After technology translation, a circuit like the one in Figure 6.3d is obtained. At this point, the specific location of the elements and the specific connecting paths (wires and switching logic) to be used have not been determined yet. These issues are addressed in the placement and routing processes, described next.

6.2.3.3 Placement and Routing

The placement and routing processes are in charge of providing fully mapped logic and fully specified interconnects, respectively. In the placement process, the already device-specific circuit must be mapped into specific locations within the configurable fabric such that it meets two basic, but opposite, criteria: Connected elements must be as close as possible in order to minimize signal propagation delays, whereas logic density (or occupation) should be not too high in order to make routing of all required connections possible with the available routing resources. Typically, placement and routing follow an iterative scheme, such that (preliminary) placements are followed by (nondetailed) routings. In each iteration, delay optimization steps are executed in order for critical delays to be minimized, and routing is progressively defined (in terms of percentage of routed signals). If there is no routing solution for some resources within an area, some elements are swapped, displaced, or separated in order to facilitate routing. Routes that were not feasible in the previous iteration are normally routed first, assuming that the easier ones will still be routable afterward.

Placement and routing are among the most time-consuming tasks in the design flow. In spite of the simplified explanation given earlier, the actual procedures are very complex indeed. For instance, the possibility of stagnation of the iteration loop is addressed not only at design tool level but also at architecture level. As discussed in Section 2.3.3, FPGAs include local connections and midlength connections of different lengths, allowing signals to reach other areas of the device using different wire lengths, letting signals avoid wire congestions while covering large distances in the FPGA without crossing too many interconnection switches (which are responsible for the most part of signal propagation delays). Design tools must also make a consistent use of global lines (also discussed in Section 2.3.3). These are mostly dedicated to clock or reset signals, but they may also be used for other tasks, such as to globally enable or disable large portions of a circuit by means of an enable signal with high fan-out. Because of the importance and complexity of this process, routing tools have become one of the key elements of the implementation tool flow.

In principle, placement is arbitrarily defined for most of the logic elements, with some exceptions. First, designers have to specify the mapping of signals to pins in the FPGA. This is done by a collection of the so-called placement restrictions (stored in a "restrictions file"), which specify the I/O type and I/O pin for each signal in the design. In addition to these mandatory placement restrictions, further ones may optionally be used to "guide" the tool in placing design components or elements in specific regions of the device. This facilitates not only more optimized designs to be obtained but also incremental or difference-oriented design to be performed. Since complex designs require a lot of computation time for placement and routing, parts of a design that have been previously validated may be consistently kept placed in the same regions with the same routing so that placement and routing efforts are mainly concentrated on the parts of the design still being built and debugged, thus reducing the overall design effort.

Once the routing process is completed, the full circuit is known, so detailed reports may be produced, as shown in Figure 6.1. On one side, utilization of logic resources is summarized, both in general terms and for every particular component (LUTs, flip-flops, slices, LBs, I/O blocks, DSP blocks, RAM blocks, etc.). The timing report is of particular interest because it includes an analysis of the critical paths and the subsequent maximum operating frequency that may be achieved. If a minimum operating frequency is specified at synthesis time, the fulfillment of this requirement is checked, and, as a result, either the slack time that is left is specified (allowing to determine how much faster than required the circuit would be able to operate) or a list of paths whose propagation delays exceed the maximum allowed time is provided. In some cases, these delays do not correspond to the most realistic situations since they may represent conditions unlikely to happen during normal operation. In all other cases, the circuit must be redesigned using time reduction techniques, such as pipelining or segmentation of large combinational areas in the design. For complex circuits with critical timing issues, this iterative design flow may be tedious and it may be the case that no solution can be found for a given design to be implemented in a given FPGA. When this happens, the first solution would be to look for equivalent devices with higher speed grades (i.e., faster), which unfortunately will most likely be more expensive. In the worst cases, the FPGA device or family must be changed, which may have significant negative implications, such as the need for PCB redesign. Therefore, it is highly advisable, in particular for time-critical designs or those where FPGA utilization is high, not to proceed to any subsequent system design step until placement and routing has been completed, and extensive timing simulations have been carried out, so that the circuit is known to exhibit correct behavior and to fit in the target device.

Figure 6.3e shows an illustrative example of a circuit after placement, where elements are placed in specific positions of the device, typically represented by their Cartesian coordinates. Figure 6.3f presents the final result after routing; that is, the circuit resulting from its description has been synthesized from a set of design specification files, then mapped and translated into the corresponding FPGA technology, and eventually placed and routed.

6.2.3.4 Bitstream Generation

Once the circuit is fully implemented after the placement and routing steps have been successfully completed, the bitstream that will be downloaded into the FPGA to configure it can be generated. This bitstream contains the data to be written to the FPGA configuration memory for the required elements inside the device to be adequately arranged for it to operate as specified.

Although the way the mapping between the configuration memory elements and the corresponding logic elements in the reconfigurable fabric is done is kept confidential by many FPGA vendors, some information is usually disclosed about the relationship between placement (in the fabric) and addressing (in the memory), allowing block relocation in a reconfigurable platform (as described in Section 8.2), or to apply memory fault diagnosis and correction schemes to critical areas. Fault diagnosis is of paramount importance for FPGAs working in environments prone to cause memory bit flips (known in the specific jargon as "single-event upsets," SEUs). For SRAM-based FPGAs configured at boot time from an external nonvolatile memory, the occurrence of a bit flip in the configuration memory can be periodically checked during operation by comparing the contents of the nonvolatile memory and the internal configuration memory. This may be combined as well with error verification and correction methods available in some FPGAs. Placement with area restrictions and the aforementioned knowledge of the bitstream structure enable this verification to be concentrated on critical areas of the design.

Bitstreams are clearly a potential source for IP vulnerability, because a design might be reproduced in principle by any third party having access to the bitstream. With the increasing complexity of FPGAs and the variety of programming and read-back possibilities available, this problem requires special attention. In order to mitigate it, vendors provide bitstream cyphering capabilities, the option of avoiding configuration memory read-back after programming, and other sophisticated techniques. The provision of remote configurations in networked systems is also a possibility to consider, since in this case there is no physical device (i.e., a flash memory) whose content may be copied, but instead an encrypted bitstream is transmitted every time a device is to be configured. This has the advantage of increased system maintainability (also enabling upgrades), but also the risk of malfunction in case of network failure. As an alternative approach, some FPGAs have unique identifiers (one identifier per device), allowing designs to be only used in the device with the right identifier. That is, the same bitstream downloaded to an identical FPGA device with identical PCB design will not work. Regarding security, however, as in any other technologies, there is no infallible solution in the FPGA domain either. If IP protection and design privacy are important issues, designers should contact FPGA vendors for the assessment of the specific capabilities of their devices in this context.

Bitstreams may be compressed in order to minimize memory utilization for bitstream storage, as well as to reduce reconfiguration time, which is mostly due to the process of receiving the bitstreams through the configuration port rather than due to the internal reconfiguration process itself (analyzed in Chapter 8). Simple encodings, such as run-length encoding, can be used for bitstream compression. In this type of encoding, bytes (or words) with the same value located consecutively in the bitstream are compressed by specifying the value and the number of times it is repeated. As a clear example of the usefulness of this technique, one may think on the great savings (in terms of bitstream storage and configuration time) associated with unused areas of the FPGA. While tools encode information at bitstream generation time, FPGAs are required to have the corresponding decoding resources that, fortunately, are really simple and silicon-inexpensive.

6.3 Design of SoPC Systems

The implementation of SoPC systems involves tasks related to both software and hardware development. At pure hardware design level, the main challenge for the design tools is their ability to efficiently integrate IP blocks, providing consistent platform-based design techniques and compatible IP libraries for reuse and rapid integration. Consistent methods are also needed for all other aspects of the whole system design flow, so synthesis-, simulation-, and verification-related aspects must also be considered for an efficient integration. From the software point of view, the main problems are associated with debugging the whole software system (possibly made up of a -real time- OS, drivers, and custom software) when implemented in real hardware.

There are tools that support the design based on embedded soft or hard processor cores (described in Sections 3.2 and 3.3, respectively). In general, they include peripheral IPs, soft processor customization tools, software development tools, hardware/software debuggers, hardware verification tools, libraries, and software and hardware code download methods to development boards.

Additionally, SoPC systems can be complemented with standard modules that are generated from core generator tools, embedded in the design environments, capable of generating instances of components such as memories, interfaces, encoders and decoders to/from multiple standards, or arithmetic operators, among others.

In the simplest cases, the SoPC can be directly created without the need for any additional HDL description. Designers just have to choose the processor core, configure its features in some kind of graphical user interface, define the types and sizes of memories, and select the peripherals to be included (or import custom peripherals described in HDL). Once the hardware structure has been defined, programming can be carried out, usually in high-level languages. After programming is complete, the SoPC system can be generated and the FPGA configured.

Hardware and software design tools for SoPCs are analyzed in Sections 6.3.1 and 6.3.2, respectively, whereas core libraries and core generator tools are addressed in Section 6.3.3.

6.3.1 Hardware Design Tools for SoPCs

As described in Chapter 3, there are many diverse hardware elements a SoPC system may consist of. However, without loss of generality, they may be divided into two categories, namely, interconnect masters (usually microcontrollers or DMA blocks) and slaves (such as specialized peripherals, hardware accelerators, or memories), arranged in single- or multimaster architectures, and either memory mapped or streamed (i.e., containing addresses and data connections or just passing data between data producers and data consumers—usually through FIFOs, respectively).

In order to better understand how tools help in the development of SoPC systems, it is also worth recalling that some of these constituent elements are hard cores (prefabricated in the silicon), whereas others are soft cores (built from the resources of the reconfigurable fabric). Regarding hard cores, two possible configurations exist in SoPC systems:

1. Coarse-grain configuration, where the components taking part in the design are activated and all others remain inactive.
2. Conventional FPGA configuration, where elements attached to the interconnections between the hardware part (processing system) and the configurable part (the configurable logic section) are set.

The structure in Figure 6.4 (that may be found in devices such as Xilinx Zynq or Altera Cyclone V) contains both types of configuration mechanisms. On the one side, the processing subsystem, which may include just one or several coupled processing cores, is equipped with a huge variety of "standard" peripherals, such as serial (e.g., I²C, SPI, or UART) or network interfaces (e.g., Ethernet or CAN interfaces), or GPIO. Each one of these may come with a variable number of elements, which may or may not be needed for a particular application and, therefore, may be selected to operate or be left inactive. For instance, the external inputs and outputs of the interfaces can be directed to virtually any pin in the FPGA by means of an I/O MUX. This corresponds to coarse-grain reconfiguration since the user decides whether to use a whole block, not to use it, or use another one instead.

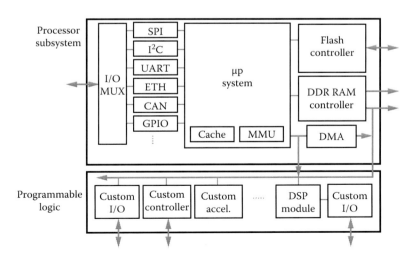

FIGURE 6.4
Simplified example of an SoPC implementation on a dual-core hard processor platform.

When commercial development boards are used, the connectors or external circuits to which the FPGA chip is attached to are predefined, so the connectivity that the I/O MUX must provide is "suggested" for each particular board once the customization definition files for that board are installed in the design environment. This relieves designers of the need for customizing external connectivity, so their only task is to "click and select" every required individual module.

On the other side, the customization enabled by the configurable logic subsystem (the FPGA part of the SoPC) provides much wider adaptation possibilities. The variety of elements available to make up a complete, application-tailored system allows designers to include, among others, the following:

- Custom I/O peripherals.
- Standard I/O peripherals not available in the processing subsystem. This includes the fact that additional ones may be required when the number of peripherals of a given type available in the processing subsystem is not enough for the target application.
- Custom controllers, which operate almost independently of the programs running in the processing part, but which can be configured to some extent by means of registers accessible by the processors.
- Hardware accelerators (either custom or standard), which can be used as replacement for repetitive software tasks to increase processing speed. In this case, there is no FPGA I/O connectivity associated since connections are directly made between the processing and configurable parts of the device.
- I/O peripherals with strong processing requirements (e.g., filters), built from DSP modules.
- Bridges to standard connection interfaces different from the ones used natively in the SoPC, in order to reuse existing modules from other IP libraries.
- Bridges to support hierarchical communication structures, in order to separate traffic at different levels of interaction, for instance, as discussed in Section 3.5.
- DMA controllers, which produce specific transactions to/from other elements, in order to achieve high data bandwidth between two ele-.ments (e.g., a memory and a high-speed I/O) without intervention of the processor core(s).
- Of course, soft processors. The use of several processors allows software tasks to be more efficiently distributed, or task execution to be isolated.

All these elements may be important for the design of different embedded systems. Examples of the use of FPGAs in industrial embedded systems and the resources that make these devices suitable for different applications in the area are described in Chapter 9.

One of the main challenges of SoPC design tools is the need to deal with such a huge variety of elements and use generic ways to connect them, avoiding complex, time-consuming, and error-prone designer tasks. The generalization of connectivity options is achieved to some extent by classifying connections according to several different concepts (most of them analyzed in Section 3.5) as follows:

- Memory-mapped or streamed connections.
- In the case of memory-mapped connections, interfaces can be either master or slave, and there may be a single master or multiple masters.
- Low- or high-speed connections, supporting single or burst transactions.
- External or internal connectivity. SoPC tools allow internal modules to be easily interconnected through compatible interfaces (either memory mapped or streamed). It is also possible to connect internal modules with others elsewhere out of the SoPC so that they have access to other parts of the design implemented in the FPGA or to external devices. In the first case, the SoPC design tools provide specific ports in the SoPC subsystem entity, which allow it to be connected to other elements in the device, designed using other approaches. In the case of external signals, the tools also provide the ports required to communicate outside the SoPC, but taking into account that they have to be connected to FPGA I/O pins, placement restrictions for I/O connections must be provided, as discussed in Section 6.2.1.

It is advisable to make efforts for designing IP modules that guarantee compatibility with standardized connections. Although at first it may seem that designing them "by hand" provides more control to the designer and simplifies the design process, actually there are many clear advantages of making them compatible with SoPC standardized interfaces and designing modules as library components that can be integrated in not just a given one but many SoPC designs. In other words, companies or design teams should consider as a fundamental strategy the development of libraries of proprietary compatible IP modules, which can later be seamlessly integrated in multiple SoPC designs. By having access to such libraries, SoPC designers may simply select IP modules from the library, attach them to the SoPC under construction, and connect them to other elements using standardized interfaces. Moreover, interface standardization allows the design tools to verify whether connections are defined in such a way that they will operate correctly.

This approach is also compatible with design parameterization, for instance, using "generics," a VHDL feature that enables hardware customization according to generic parameters specified at design time (i.e., statically assigned case by case). Using this approach, whenever a module from an IP library is selected to be attached to an SoPC system under construction, the specification of the corresponding design parameters is required from the designer by an automatically generated tool interface.

As can be seen, connectivity standardization and classification allow SoPC designs to be arranged such that all interconnections among internal elements can be made in a controlled way. This ensures, on the one side, correct system behavior and, on the other side, the possibility of building a layered structure of software elements on top of the hardware, which facilitates software integration. For instance, memory-mapped interfaces with multiple slave elements sharing the same memory map are automatically assigned memory positions and memory ranges, and the corresponding memory decoding logic is automatically attached to the design. Additionally, all memory addresses are supplied as constants with automatically assigned names, which allows software programmers to more easily handle the associated elements (registers and memory-mapped areas), providing an abstraction layer that "isolates" hardware and software designs, so that their respective developments can be overlapped. Software designers do not need to know the actual memory addresses in the final hardware design but can use the automatically generated names instead. This separation also grants code portability; that is, software controllers and drivers may be reused regardless of the memory assignments done in the automated hardware design. These issues are further explained in Section 6.3.2.

6.3.2 Software Design Tools for SoPCs

SoPCs require software to run on the processor (or processors) so that the combined action of these programs and the tasks performed by the various hardware elements included in the SoPC system produce the expected results. Hardware and software designs should be carried out from a partition of tasks that, considering available resources as a restriction, allocates most computing-intensive tasks to hardware elements and least frequently used tasks, control-based tasks, or (likely) high-level tasks to the software part.

Considering the many different configuration possibilities, the complexity of the underlying architecture can be excessive for the software designer to handle. So, for tools to ease rapid deployment of hardware/software solutions, they must provide sufficient levels of abstraction as well as automation processes that enable designers to easily do two things: verify the integration of software and hardware, and develop and test the necessary control software.

For complex stand-alone processors, a programmer's model is provided where registers and main addressable components (i.e., the elements to be handled from the software designer's perspective) are contained. Similarly, tools for software development in SoPCs must provide such levels of

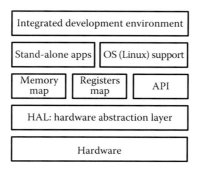

FIGURE 6.5
Layered structure for software development support.

abstraction. Figure 6.5 shows a simplified layered structure to connect hardware and software, upon which these tools rely.

Moving bottom-up the hardware representation, including all elements that are actually placed in the FPGA, is made abstract by a layer usually known as hardware abstraction layer (HAL). This simplifies the representation and hides pure hardware details, just keeping the elements that may directly or indirectly be used by upper layers.

Although design tools are more and more integrated, for instance, through the use of standardized programming environments (such as Eclipse) for programming and verification, there is still a frontier between hardware and software design tools. The HAL layer provides the information to be passed from the hardware design tools to the software design environment.

From the HAL, memory and register maps are obtained, allowing software designer to easily address any element in the hardware that can be accessed from the processor(s). Every such element is given a constant definition that matches the actual address in the hardware. The name of the constant is normally obtained by the concatenation of at least the bus name and its associated memory map name (automatically or user provided), the device name (normally user provided), and the register generic name (provided by the tools). Although using these names for developing programs might result in being tedious, and even in some cases error prone, it ensures portability (in turn improving productivity): Modifications in the hardware forcing a small redesign of a block within the underlying structure will not affect other blocks' related software code. For addressed memory areas, constant names are also used to specify upper and lower limits, or ranges. In addition to these "canonical" names, user-defined aliases can be used for easier identification of specific registers (e.g., the most commonly used) in the design.

Many of the aforementioned registers are required to be used during the initialization phase of the system for adequate configuration setting. To facilitate this, tools also include APIs with functions automating the initialization, control, and access processes. In some cases, functions for component verification, benchmarking, and validation are also available. Hardware verification and rapid application deployment are enabled by these high-level functions.

At higher abstraction levels, a distinction must be made between bare-metal applications and those handled by an OS (as also discussed in Section 3.1.1.2). Bare-metal applications rely on functions to access hardware resources and on the abstraction levels provided by the HAL and upper layers. For applications managed by an OS, there is another abstraction provided by the drivers supporting the underlying hardware. The use of drivers improves portability, but on the other hand, the development of the ported embedded system is more complex because of the need for also porting the OS. However, tool and FPGA vendors are aware of this complexity, so they provide customizable porting, either directly or through agreements with the companies supporting the OS. There is also a need for porting or integrating other components, such as filesystems on top of flash memory controllers so that the OS can easily handle all information. These flash memories need to be partitioned in order to allocate bootloaders and root filesystems. This process is sometimes cumbersome because we are dealing with embedded systems, so there may be a lack of terminals to interface with the system.

Once these steps are completed, the advantages of using an OS will become apparent. For instance, some standard peripherals can be configured automatically or by specifying the required configuration in the initialization files of the OS. For a very usual feature such as Ethernet communication, module detection, initialization (including link detection), or speed negotiation are automatically carried out and network and transport protocols are provided by the OS. Therefore, software designers just need to develop an application that relies on OS sockets, using libraries similar to those available in conventional platforms such as personal computers. In the case of bare-metal applications, the TCP/IP or UDP/IP support provided by the OS is replaced by somehow equivalent high-level functions. Lightweight implementations of this part of the communications stack may be provided to relieve users of the need for configuring the access to the physical device outside of the FPGA and the media access control layer inside the FPGA. The latter can be available either in the configurable fabric or, in many SoPC devices, as a hardware block.

The top software development support layer in Figure 6.5 is the integrated environment itself. As mentioned before, these are standard environments (with some customization), such as Eclipse, separated from the hardware development environments. They integrate tasks such as the consistent management of all data related to software development, configuration management, software project management, interfaces with integrated compilers, deployment tools (interfaces with bitstream downloaders and executable code downloaders), execution control, access for debugging and profiling control, and its associated windows for visualization of results.

Software applications are packed into projects with support for more than one platform. Projects also integrate source code, compiled libraries (static and dynamic), intermediate representations (i.e., assembly code), and configuration data. Intermediate files are checked for consistency with respect to other elements in the project, providing automatic recompilation of the code when some portions they depend on are changed. For user-defined information, such as source code, version control may be activated, allowing programmers to keep control of the changes, backtrack to former versions, or issue versions and releases of the code.

Compiler options may be selected either manually (by expert users) or from a list of options in wizards or graphic windows. Compilation may be launched as an individual operation by the user, or automatically in response to changes in the code they depend on. After successful compilation and library linking, the executable code is ready to be downloaded to the FPGA, either together with the bitstream or separately.

During development and verification, execution may be controlled from the integrated environment, once the executable files have been downloaded to the FPGA. For bare-metal applications, this may be done in two ways:

1. By installing a bootloader that provides a communication path between the FPGA and the computer hosting the integrated environment
2. By using dedicated logic, usually associated with the JTAG programming interface, with some possible extensions for debugging, which enable memory and register content download and execution control

When using an OS, execution control is achieved by remote debugging, debugger accesses being carried out through either a serial communication link or a communication port set on top of the TCP/IP stack, if available. Debugging tools for software validation, profiling, and mixed hardware/software verification will be specifically analyzed in Section 6.6.

After the validation of the hardware and software parts, and of the correct interoperability among them and the environment of the embedded system, final deployment involves permanently setting all required information, that is, bitstream, executable file(s), and, if it exists, OS image. All this information can be included in a single file, stored in a nonvolatile memory, from which the FPGA will be configured at power-up, as described in Section 6.2.

6.3.3 Core Libraries and Core Generation Tools

Apart from the tools mentioned in Sections 6.3.1 and 6.3.2, there exist tools intended to generate standardized, but to some extent customizable, cores (e.g., memories) that may be integrated in a design in the same way a custom-designed block may be. These tools produce HDL or netlist-level models of

the core. Some core generators are based on highly configurable and parameterizable models that, by means of using VHDL "generics" or constants
included in customizable packages, give designers the possibility of rapidly
obtaining a core that fits their needs.

Many of these cores are for free use, whereas others are only usable
at some cost. The latter are usually from third parties whose business is
based on selling IP libraries. The ways they are licensed differ, but in some
cases, designers are allowed to use them during a given time (they would
stop operating after some hours of use, just giving the chance to test their
usefulness before purchasing them) or to use them while there is a JTAG
cable connected between the development board and the integrated tool
environment.

Some cores are technology independent, in the sense that they may be
used in more than one FPGA technology, whereas in other cases, they are
family or even device specific.

In the case of standardized communication interfaces, memory interfaces, and other I/O-related standards, especially when they are associated
with specific devices, the cores include specifications about location, I/O
type, and timing constraints in order to ensure that they will work properly on the target device. Also, when standard internal connectivity (such
as a slave interface for a given task) is included, the cores are conveniently
integrated into the SoPC tools so that connections can be automatically
made in a consistent way. For standard interfaces and functions, certified
modules are provided in many cases, ensuring full compatibility with the
related standards.

Vendors and associated third-party partners offer huge libraries of components, which include far too many cores to be listed here. Therefore, just
some significant examples are given here:

- Different types of memories, such as single- or dual-port RAM memories, or FIFO memories, among others. They can be parameterized
  in size (data width, number of addresses) and have either synchronous or asynchronous access. FIFO memories may be equipped with
  different access and control interfaces (e.g., configurable half-full or
  half-empty flags). According to their requirements and the resources
  available in the target device, memory cores can be adapted to be
  implemented using embedded memory blocks or distributed logic,
  as discussed in Section 2.4.2.

- Memory controllers and interfaces to external memories, such as all
  kinds of compatible DRAM (DDR, LP-DDR, etc.), flash controllers, or
  DMA controllers.

- DSP-related blocks, such as digital filters, modulators, and demodulators for signals under various standards, compressors and decompres-
  sors, cyphering blocks, FFT, DCT (and their inverse operators), tools for

designing linear algebra blocks and related blocks, CORDIC blocks for trigonometric math, and signal synthesizers (e.g., sinusoidal generators).

- A wide variety of communication interfaces, certified in many cases, targeting different applications. These may also come as a set of modules allowing custom (application-dependent) communication interfaces to be built by combining blocks such as demodulators, data scramblers, interleavers, or error-correction encoders/decoders.

- Modules whose application domains are very specific (and so are their architectures and features) but widely used. For instance, there is a large variety of video signal compression and decompression blocks for specific standards, such as H.264 families. There are blocks targeting applications in the aerospace sector (compatible with ARINC or NIST standards, for instance) as well as in industrial or medical domains. On many occasions, these specific cores are available subject to some kind of licensing scheme.

6.4 HLS Tools

As introduced in Section 6.1, HLS tools offer the possibility of mapping algorithms into hardware from descriptions that are not time explicit or, in other words, do not contain information about transactions between registers in every clock cycle. These tools make the appropriate scheduling of operations in a given set of operators, which may be reused over time for different purposes.

The algorithm specification defines relationships between variables containing data and operations, so data are transformed along the algorithm execution. The tools identify involved operators and data dependencies, in order for the modules in charge of these operations to be reused at different times of the algorithm execution. They also generate the multiplexing schemes and the associated control required to select the appropriate data path(s) at every time point during execution. As a result, they provide

- A data flow graph, which contains the registers required to hold data, the multiplexing schemes required to feed operators with these data, and the operators themselves

- A control state machine, which controls the data flow graph in order for the required operations to be performed in the required sequences

Reuse of operators can be maximized to reduce logic resource utilization, usually at the expense of longer latency. Alternatively, if execution speed has higher priority than size, the circuit may be "widened" by multiple instantiations of operators so that parallelism may be exploited. In this case, pipelining, loop unrolling, parallel memory access (memory reshaping), and I/O adaptation are the main techniques used to speed up algorithm execution. These techniques are briefly described here:

- Pipelined structures achieve high execution speeds at the expense of high number of registers and long latencies. A well-designed pipelined circuit should have all stages performing operations and holding data, cycle by cycle, with data coming from various execution cycles, as long as the signals are being propagated by the pipeline. Thus, pipelined structures are incompatible with resource reuse since structural hazards would be produced.

- Loop unrolling is a technique that uses several functional instances for the inner loops of the code so that all iterations within the loop are executed in parallel. In order for this to be feasible, the loop must contain a fixed, predefined number of iterations (i.e., it does not depend on a variable but on a constant). If loops are nested, more than one loop may be set to be unrolled, but the chances for huge resource utilization increase. In general, this technique requires high resource utilization but few additional registers, and it should be complemented with memory reshaping and I/O adaptation, because all resources must be fed with the appropriate data at high speeds and simultaneously, otherwise, no performance improvement would be achieved.

- Fast access to memories by the functional resources is crucial to achieve high computing bandwidth. With this purpose, memories (in particular those storing vectors or arrays) may be set to use wide parallel buses, capable of providing data to the possibly replicated computing resources at the required speeds. Since memory contents are the same, memory utilization inside the FPGA remains unchanged and the only overhead is that caused by parallel wiring. For this reason, this technique is called memory reshaping.

- Data from the external elements have to be fed to the blocks designed through HLS techniques fast enough for all required data to be available at the right times. Similarly, these blocks must be capable of delivering output values to their destinations at the right times. High data throughput may be achieved by using DMA engines on dedicated ports. They may be embedded into the system under design for the control flow part to produce the proper transactions at the right times. Apart from traditional HLS tools, which are being integrated into design suites, there are also tools aimed at embedding (in a somewhat automated way) hardware accelerators within SoPC systems. They are targeted to a restricted set of devices or families and are conceived to support software designers with little expertise in hardware development.

A special case of this approach is the development of hardware accelerators from programming languages that allow explicit parallelism to be described. OpenCL is becoming a widely used standard for such specifications because of its adequacy to cater to a variety of devices, such as GPGPUs, multicore systems, or SoPCs. It also supports heterogeneous computing, in the sense that different portions of the code may run on different computing platforms, as discussed in Section 3.1.1.1. This is very convenient for the newest FPGA families, which integrate several different hard processing fabrics in the same device. Because of its expected increased significance, the issues related to the design of these particular accelerators are discussed in Section 6.5.

6.5 Design of HPC Multithread Accelerators

As analyzed in Section 6.4, HLS synthesis tools can be a good performance booster for accelerating certain critical tasks within a control system with limited extra design effort. An alternative and, in many cases, advantageous solution is to define parallelism in an explicit manner in the source code of the algorithm to be accelerated. Languages with explicit parallelism, such as CUDA or OpenCL, share the same model of computation, that is, the way efficient code is to be produced to achieve significant acceleration. While CUDA is specific to NVDIA GPUs, OpenCL is becoming a *de facto* standard due to its portability to different platforms, such as multiprocessors, GPUs (and GPU clusters), FPGAs, or even heterogeneous systems formed by combinations of these platforms.

OpenCL ensures code portability between different computing devices, although performance is not guaranteed. It is clear that the computation model underneath the code and the hardware architecture on which it is executed play a crucial role in the resulting performance. As a matter of fact, if the code is not written carefully enough, performance can be degraded to the extent that it can be worse than that achieved using a single processor.

The computation model provided by this type of languages relies on a multithread approach, based on the parallel execution of multiple basic elements (called work items in OpenCL and threads in CUDA), with different levels of interaction between them. Each work item/thread has its own private memory for independent computing, ensuring that the maximum possible bandwidth is achieved.

Work items/threads can be bundled into work-groups/thread blocks. All bundled elements share a second memory level, called local memory, which can be accessed by any of them, but with some restrictions. Local memory is multibank, so it has multiple ports for parallel access from all elements at the same time. However, each memory bank can only be accessed by one of them at a time, except if access is gained from the same memory position within the bank. Special care must be taken with this type of accesses since good parallelism exploitation comes from parallel coalesced accesses to this memory, with no congestion due to chaotic access.

Each work-group/thread block is expected to be fully executed in the same computing unit (CU), but since the computing models dictate the execution of different work-groups/thread blocks to be independent, each of them may run in a different CU. All work-groups to be executed are bundled into a kernel. A kernel is invoked whenever there is a need to perform multiple work items/threads in parallel. If the number of work-groups to be executed is higher than the number of CUs available, execution is sequenced until that of all work-groups/thread blocks in the kernel is finished.

All work-groups/thread blocks in a kernel also share a third memory level, called global memory, which is accessed by them through burst transactions, in order for data throughput at this level to be maximized. Every work-group/thread block and work item/thread has a numeric identifier that enables each of them to access their own sets of data in this memory. These identifiers can be one, two, or three dimensional in order for different data organization to be possible, allowing algorithm memory needs and work-group/thread block and work item/thread organizations to match. For instance, one-dimensional partitioning is adequate for dealing with single signals, whereas two-dimensional (2D) access is more efficient for 2D image processing, and three-level identifiers are the best solution for finite-element analysis (e.g., mechanical) of a 3D structure.

Kernels are invoked from a host, which executes serial code containing kernel invocations. Kernels are then executed on the so-called device, which contains the CUs required to accelerate kernel execution. Host and device have their respective own memories, so memory transactions are required between them for data provision and result collection. If kernels are not too computing intensive, the time saved in parallel computing may be counteracted by the time used in memory transactions. Another possible cause of performance degradation is the need for synchronization between parallel work items/threads, equivalent to a barrier in multiprocessing terminology, which might cause some CUs to be underutilized.

A host program may, of course, invoke more than one kernel along its serial execution. It is also possible to specify how many accelerators (CUs) are to be allocated in the FPGA for each kernel. Additionally, it is possible to modify the logic inside the FPGA by partially reconfiguring the area devoted to the accelerators so that different combinations of CUs can be used along host program execution. An example of the use of partial reconfiguration for this purpose is described in Section 8.3.3.

The main FPGA vendors offer tools for accelerating OpenCL kernels implemented in FPGAs. They are suited to be used with powerful high-end FPGA boards (the devices) hosted in personal computers (the hosts) and connected through PCIe interfaces. It is also possible to run the same OpenCL programs in the host processor to verify functionality, verify them with hardware simulators, test them in real hardware using just one CU in the FPGA, or fully verify them.

Although multithread acceleration tools are intended to support software designers, some knowledge of hardware acceleration and, more importantly, the characteristics of the computing model and their impact on acceleration are required. CUs capable of executing one work-group/thread block at a time are obtained by means of an HLS synthesis process, but specific directives (or pragma declarations) are required to customize the number of work items/threads per work-group/thread block, which in turn determines the size of every CU, as well as the number of CUs to be allocated in the FPGA fabric.

Same as for stand-alone HLS tools, OpenCL acceleration environments offer estimation tools in order to explore the design space (basically area and performance) before going into the detailed design process, which is quite time-consuming. Estimations may be obtained about latency in every CU, throughput, and resource utilization of each CU, among others.

6.6 Debugging and Other Auxiliary Tools

Most tools and methodologies described in previous sections are related more with the design of FPGA-based systems than with their validation, with the exception of simulation tools. Simulation is an essential part of the whole development process of FPGA-based systems (actually of any electronic system). There are other features that, not being strictly required such as simulations, are very useful to identify design problems, particularly in the case of complex systems. These optional tools for hardware and software debugging are described in Section 6.6.1. Other auxiliary tools that facilitate the design of FPGA systems, such as pin planners or power estimators, are addressed in Section 6.6.2.

6.6.1 Hardware/Software Debugging for SoPC Systems

The complexity of SoPCs makes it necessary to have tools available not only for software debugging (like in any other microprocessor system) but also for verifying the interaction between the software and hardware parts. Actions taken by the programs running in the processor cores have an impact on the hardware, which is not always easy to follow and verify just with the aid of external signals. Interfaces between elements may not be exercised properly, making it difficult to check whether or not the operation of the system will be correct in all scenarios.

When problems arise, it may be extremely difficult to tell if they come from an inconsistent software design, an incorrect interfacing, or any other reasons. For "small" designs, where there are enough pins left in the FPGA, debugging actions may be performed by taking outside of the chip critical signals that provide indications of correct or incorrect behavior of the system. However, this method is usually difficult to implement and is error prone since it requires navigating through the system hierarchy of modules so that the target signals can be identified and taken out from (possibly deeply) embedded blocks.

On the other hand, it may be not possible to correctly transfer fast-switching internal signals outside of the device, for instance, because of signal integrity problems or because available pins cannot achieve the same switching speed.

Finally, tracing and fixing potential problems in a conventional manner might require the design to be modified a high number of times, which implies the need to repeatedly run the time-consuming synthesis process; whereas trying to identify potential problems beforehand and having embedded instruments selecting and analyzing the signals of interest at runtime would help in reducing the number of synthesis processes required for a successful design to be completed.

Therefore, it is very interesting to have specific debugging tools targeting SoPC systems, allowing instrumentation to be embedded inside the devices, so that internal signals can be directly monitored in place. Fortunately, it is currently possible to have very powerful embedded instrumentation implemented inside FPGAs, whose features are in some cases comparable to those of conventional instruments, such as logic analyzers, and which are capable of combining software and hardware debugging.

It is important to clarify that, since many SoPC systems may have real-time constraints, the addition of debugging resources should neither have any impact on execution time nor use resources that would be required for normal system operation. The main debug mechanism to access resources inside the FPGA in a nonintrusive way is the use of the JTAG interface, which is not typically used during normal operation of the device. Remote debugging may also be possible by means of a TCP/IP connection to the system, for instance, when an OS is used.

6.6.1.1 Software Debugging

The main tasks involved in software debugging are the inspection of the contents of memories and register banks, and execution control, which relies on step-by-step execution, breakpoints, and watchpoints. For embedded hard processors, the logic resources required to provide such features are embedded into the logic, so there is no need to specify them as part of the design. In soft processors, where it may be necessary to specify their inclusion, integrated design tools provide mechanisms to attach the required embedded logic to the processor core.

Software debugging is performed from the software development environment, where it is seamlessly integrated. For instance, breakpoints and watchpoints can be directly inserted in the source code (or, rarely, in the assembly code). There are no significant differences with respect to conventional software debuggers, one of them being the potential need for controlling more than one processor execution at a time, given the fact that the system may include several processor cores. In this case, a breakpoint or watchpoint may be used to stop execution in just the processor where they are inserted or in all of them. More advanced synchronization techniques are also available, but they require the insertion of additional logic, as discussed in Section 6.6.1.2.

The way breakpoints operate on the underlying hardware is by continuously comparing the contents of the program counter with the address in the instruction memory where the breakpoint is placed. This can also be done by checking the address bus in the instruction memory. Access must be gained in the bus between the processor and the cache memories. Otherwise, if comparisons are done close to the program memory itself, instruction cache must be stopped to avoid hit accesses not observable in the outer bus.

Watchpoints are similar to breakpoints, except that they check for data memory instead of instruction memory addresses. Thus, the same considerations apply to data caches and watchpoints as to instruction caches and breakpoints. In some cases, the condition to stop execution is based on a specific value being read from or written to a data memory position. In such cases, a second comparator is required in the data bus of the data memory, in addition to the one checking addresses.

Performance monitors are also commonly available, providing a method to measure execution time, bus utilization, and so on. They allow profiling execution of software in the real system, with the aim of optimizing resource utilization.

6.6.1.2 Hardware Debugging

Hardware debugging is not as specific as software debugging. Its features are related to the ability to observe internal signal activity in the hardware by means of generic, but configurable, embedded instrumentation. These instruments internal FPGA resources, so their use can be a problem in systems with high resource utilization. In particular, in order to be able to acquire the activity of signals at their nominal speeds, internal memory resources (usually relatively scarce) may be necessary to store the corresponding traces, so special care must be taken when specifying the amount of memory reserved for such task.

Before synthesis, the tools allow users to specify the signals to be monitored after the system is designed and implemented in the FPGA. These signals are kept in the design-along stages, preventing them from disappearing due to circuit optimizations during the design process. This is automatically accomplished by the tools, either by making simple modifications at netlist level or by including an MUX to select signals at runtime, so that they are connected to the embedded instrumentation for analysis and/or storage. The second option is more flexible, but could incur higher resource utilization overhead as well as some speed degradation, which could be neglected if operating frequency is not a critical design constraint.

Different FPGA vendors offer similar embedded debugging instrumentation, which in essence is also similar to the functionality provided by conventional logic analyzers, mainly based on the acquisition of a group of signals, synchronized by events in a sampling signal, and triggered by the occurrence of some events or combination of events in a signal or set of signals.

The sampling signal determines at which moments of the execution the observed signals are to be acquired. It may be any clock signal in the design, but it is not the only solution. For instance, if transactions in a communications link are to be analyzed, once it is known that their timing behavior is correct, any data validation signal (data strobe, an acknowledgment signal, an enable signal in an input data register, or similar) may be used as sampling signal. In this way, only actual transactions are registered, saving a lot of acquisition memory or, in other words, allowing for longer data acquisitions. In these cases, time stamps can be registered together with the acquired data in order to determine the instants when data transactions actually occurred.

Since in most cases acquisition has to be carried out under nonrepetitive conditions, specific events in trigger signals cause acquisition to start. These may be:

- *Simple trigger events*: value of a given signal (or combination of values of a group of them), edges in certain signals, or similar.
- *Advanced trigger events*: for instance, a trigger may be controlled by a counter, which activates it after a given number of occurrences of a simple trigger event.
- *Cascaded or interconnected trigger events*: sophisticated trigger conditions can be specified by different combinations of simple trigger events.

The effective moment to react to the trigger event (whatever its complexity is) can be configured to be the exact one at which the event is produced or a given number of sampling periods before or after its occurrence. For data acquisition to effectively occur before a trigger event is produced, data sampling needs to be active at all times. Acquired data are stored in a memory that operates as a circular buffer. When the memory is completely full after the trigger, acquisition is stopped.

Virtually any signal or group of signals (buses, internal control signals, or elements that are not accessible through read-back operations from any memory map) can be monitored and the corresponding data stored in memory for debugging purposes under the occurrence of any trigger event. Since the sampling signal can be specifically generated for that purpose, sampling speed is in principle only bounded by the limits of the technology.

Debugging tools allow results to be visualized in several different ways:

- The most usual one is equivalent to the visualization of simulated waveforms, with similar interfaces and analysis tools.
- Signals can be shown as monitoring windows, virtual LEDs, or the like. Virtual LEDs can also be used to virtually represent outputs, which is useful when the FPGA board being used has not enough visualization capabilities for the target application.
- Similarly, inputs can also be made virtual, using virtual switches to set the values of internal signals or registers, without having them physically available in the FPGA board.

6.6.1.3 Hardware/Software Co-Debugging

Both hardware and software debugging can be combined, mainly to verify interaction between both parts of the SoPC. For instance, a breakpoint may be used to trigger the acquisition of an embedded logic analyzer, or a trigger condition may be used to stop the execution of a program running in a processor core. With this feature, it is relatively simple, for instance, to set interrupt conditions as triggers or to verify whether interrupts are occurring when required.

6.6.2 Auxiliary Tools

6.6.2.1 Pin Planning Tools

Modern FPGAs have a huge amount of pins, and the selection of the most suitable ones to ensure the proper operation of the whole circuit the FPGA takes part in may not be a simple task. Some pin locations are preassigned for specific signals (e.g., clock signals) or specific types of signals (e.g., DRAM memory interfaces, PCIe connections, other communication transceivers, or, in general, any signal with specific timing or speed restrictions, requiring a preassigned position close to the embedded hard blocks that use or generate them). In addition, some I/O configuration possibilities require specific supply voltages in auxiliary supply pins, which may be incompatible with other I/O standards. Finally, there are also restrictions associated with noise immunity, especially for low-voltage signals, as well as some requirements that limit the number of simultaneously changing signals in order to avoid voltage drops due to excessive switching activity.

I/O pins are organized into different banks, each one with its own power and auxiliary voltage supplies (which allow, for instance, unused banks to be turned off to reduce overall power consumption). Some of the aforementioned restrictions affect each bank separately.

Since handling all these restrictions may be not a trivial task, design environments include pin planning tools. They allow signals to be assigned to pins either "manually" or from the placement restrictions file discussed in Sections 6.2.1 and 6.2.3.3 and, later, the fulfillment of the restrictions to be verified, flagging warnings, incompatibilities, or errors* that might force the pinout of the design to be changed. For instance, pin planning tools advise on the need for having differential signals close to each other to avoid noise, or even for using prespecified pin pairs for them. As another example, LVDS signals cannot have nearby signals with high switching activity, and it is recommended that they are placed near to or even surrounded by pins set to ground for noise protection. All signals in a bus are preferably (if possible) spread out through the same FPGA bank, or even split into two banks, for signal integrity reasons.

Once the pinout of the FPGA design is decided, pin planning tools should be run in order to allow the PCB design to be performed in parallel with the FPGA design.

6.6.2.2 FPGA Selection Advisory Tools

In many cases, there is no specific requirement guiding the decision about which FPGA to use in a particular application, apart from the needs for enough resources and achievable speed. Since most FPGAs are available in different speed grades, the speed issue can be addressed once the maximum operating frequency posed by the design is known.

FPGA selection advisory tools provide a means of choosing the adequate FPGA that fits the design needs. They may operate from area estimations after synthesis or through specific analyses. If there is uncertainty about the final contents and functionality of the design, it is advisable to select a range of devices from a family of pin-compatible FPGAs so that, after detailed design, the right device may be selected among a set of them, in order for the cost to be reduced.

6.6.2.3 Power Estimation Tools

Power estimation is useful during design space exploration in order for the most appropriate design among several implementation possibilities to be selected and for the maximum currents to be supplied by the power converters to be determined, in order to advance PCB design. In the first case, absolute power consumption values are not so important since relative power consumption among different approaches is the factor that may influence the selection. In the second case, only rough estimates are required since

---

* Warnings must be analyzed, incompatibilities or errors must be solved.

some security coefficient will be applied resulting in an increase in the current the power converters would be able to supply, or the use of heat dissipation techniques, and so on.

These are the reasons why power estimators are just aimed at supporting decisions during the early stages of the development of the circuit, rather than at providing a mechanism to accurately estimate the actual power that will be consumed by a circuit once it is implemented. As a consequence, some early power estimations just rely on the expected amount of I/Os and resources to be used, affected by an activity factor, which is the expected switching frequency of every signal (referring not to clock frequency but to the expected activity of each particular signal). At later stages, some more sophisticated power estimators may use simulation results to estimate activity, but it is worth noting that they are prone to offer values higher than real ones since simulations try to "compress" the whole behavior of the system into the minimum possible number of clock cycles in order for simulation time to be reduced.

References

Altera. 2013. Implementing FPGA design with the OpenCL standard. White paper WP-01173-3.0.

Cong, J., Liu, B., Neuendorffer, S., Noguera, J., Vissers, K., and Zhang, Z. 2011. High-level synthesis for FPGAs: From prototyping to deployment. *IEEE Transactions on Computer-Aided Design of Integrated Circuits and Systems*, 30:473–491.

Riesgo, T., Torroja, Y., and de la Torre, E. 1999. Design methodologies based on hardware description languages. *IEEE Transactions on Industrial Electronics*, 46:3–12.

Sangiovanni-Vincentelli, A. and Martin, G. 2001. Platform-based design and software design methodology for embedded systems. *IEEE Design & Test of Computers*, 18:23–33.

Xilinx. 2014. The Xilinx SDAccel development environment.

7

Off-Chip and In-Chip Communications for FPGA Systems

7.1 Introduction

While computing is the essence of integrated systems in general, and FPGAs in particular, it is evident that data need to be transported in order to be efficiently computed. For non-data-intensive applications, communications may be neglected, and only computing-related issues are taken into account (computing performance, computing power, computing efficiency). However, for those application that work with large amounts of data (big data applications are the most representative ones), it is clear that communications can be a key factor in the overall performance metrics. It is therefore required, for many applications, to be able to plan both computing and communication resources jointly. Moreover, it might be the case where performance is degraded because of communication overhead.

As was mentioned in Chapter 1, most systems require communicating with other external elements. The requirements of such off-chip communication are dependent on many factors: speed (both in data throughput and latency), reliability (sometimes packets of data may be lost, in other cases not), distance, environmental conditions, compatibility with networks, and so on. Therefore, there is no one solution that fits all. Although there are several classification possibilities, we have made a distinction between low-speed (Section 7.2.1) and high-speed ones (Section 7.2.2) due to their different implications in both internal and external FPGA designs. However, some distinctions between different technologies are also made, as well as between data-oriented and control-oriented communication interfaces (for the low-speed ones).

Since FPGAs integrate multiple computing resources, it is also necessary to establish some basics on how to efficiently implement communications between modules inside the chip. Although it would seem that internal communication resources are extremely versatile and there should be no problems in connecting as many elements as required, with direct connections from the places where data are produced to the places where data are required, the architectural complexity of large designs impedes this straight-forward method. For instance, resource sharing forces to concentrate or distribute data through communication structures from/to different elements, and therefore some arbitration is required. Fortunately, this is alleviated by the use of standardized interfaces that, for the purpose of module reuse, and with the help of automated tools to build complex SoPCs (see Chapter 6), let complex connection schemes to be built. This way, developing multiple buses, hierarchical buses, and bridged structures is somewhat easier. Examples of such arrangements have been addressed in Section 3.5 to describe the connectivity of embedded microprocessor cores. However, the scalability of such structures is a problem for large designs. Their performance worsens with an increase in the number of connected modules for two reasons: data access saturation, produced by an increasing number of modules willing to access the same communication resource, and maximum speed (i.e., data throughput degradation), because of the increased distances to be covered between distant modules. Although heavily pipelined structures might partially resolve this last issue, the growing complexity of large SoPC designs might lead to the need of setting up networks-on-chip (NoCs) as a shared mechanism to link a larger number of interconnected modules, providing a structure that enables simultaneous connections at the same time, with no such speed (frequency) degradation when the system scales up in size. In-chip communications are addressed in Section 7.3, covering point-to-point communications, bus-based connections, and NoCs in its three different subsections.

7.2 Off-Chip Communications

7.2.1 Low-Speed Interfaces

These communication schemes are intended for control purposes or for low-data-intensive applications, where only small amounts of data are exchanged. They mostly rely on serial interfaces with explicit low-speed clocks for easy data transfer and with cost-effective hardware, both on the transmitter and receiver sides. The amount of bits to transfer is limited, normally in the case of fixed-length transactions, and in some cases, addressing of external devices allows connecting more than one device on the same interface. These interfaces are best suited to controlling external devices, such as smart sensors or similar, where the amount of data is less. For this reason, they are also intended for controlling external devices, providing a mechanism to access internal registers to set operation modes, and collecting either small amounts of data or status flags.

I²C and SPI are the most commonly used interfaces in today's world of standardized devices, controllers, smart sensors, ADCs, DACs, etc. They offer standard wiring for many off-the-shelf devices, providing a common method to interface with many of them.

There are, in some cases, specific needs, such as very low latency, which produce yet standardized interfaces with particular requirements. It is the case of the CAN bus, extended in automotive and similar sectors, where access to the bus is granted based on the priorities of the messages to be sent.

FPGAs and FPGA design tools are ready to use such communication interfaces. SoPC devices contain—normally more than one—hardwired interfaces attached to their internal processors. Also, tools for SoPC design, even for pure FPGAs with just configurable logic, contain libraries with the required modules, accessible from embedded processors by means of standard interfaces to very rapidly customize the design and add a variety of such standardized communication modules. The design of these types of systems is quite simple since, for instance, there are numerous tutorials provided as examples to build systems with I²C or SPI interfaces.

7.2.2 High-Speed Interfaces

Contrary to what one might think, there is a tendency to solve point to point fast I/O with optimized serial interfaces more than with parallel ones. For maximum throughput, several of these serial I/O interfaces may be connected in parallel to achieve really high throughput. An example of this is the PCIe interface described in Section 2.4.4, which may use, according to the specification, different number of serial lines. Details of PCIe as well as other standards that are normally used in FPGAs are discussed in Section 2.4.4. In this section, however, the criteria on how to use them and conditions of operation and precautions when such structures are used are given.

High-speed interfaces are packet-based synchronous communication interfaces with clock recovery at the receiver end since it is not possible to transmit the clock as an additional line (as opposed to low-speed interfaces). The clock is recovered with the aid of DPLL logic or similar structures. Transceivers with the required specialized interfaces are equipped in many FPGAs, and there are families or specific devices that are equipped with many of such interfaces in order to achieve a very high aggregated data bandwidth. These specialized FPGAs are intended to be used in communication devices, such as high-speed switches and routers, where processing lots of serial I/O efficiently and at high speed can only be done through hardware, not with software.

These high-speed serial I/O interfaces are also useful in the domain of HPC, where high data throughput between several FPGAs, forming an FPGA cluster, is required. Except for specific applications with a precise connectivity, such as for pipeline-streamed applications, these clusters very often serve a general purpose, so connectivity between all FPGAs in the cluster is required. Figure 7.1 shows an example of a backplane that connects five circuits, for example, FPGAs, on a fully connected topology (all to all).

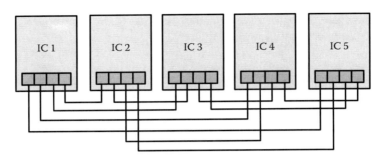

FIGURE 7.1
All-to-all backplane connection, suitable for serial high-speed I/O interfaces.

Each of these links in the figure can be built using high-speed serial I/O interfaces. These rely on differential separate pairs for transmission and reception, so, in essence, each of these connections represents four signals. Therefore, the densest part of the backplane, with six connections, would represent just 24 signals. If every pair of transmitting lines is designed to offer around 6 Gbps each, or even 50 Gbps each, which is possible in some technologies, the aggregated throughputs of the cluster would be 120 Gbps to 1000 Gbps.

The design of these serial differential I/O lines at PCB level is, however, very critical. Pins in the FPGA need to be as close as possible, and the paths along the PCB must be almost identical, with sufficient grounding around them to avoid susceptibility to noise, since they operate at low voltages, to allow for such high switching activity at pin level.

While these point-to-point connections are intended for PCB use, or for very short distance interconnections, there are cases where longer distances are required. This is the case of high-speed Ethernet connections, where internal resources in the FPGA are not sufficient to achieve the required physical layer communication requirements. In these cases, the transceivers to correctly modulate and adapt the signals to transmit through longer distances must be held outside the FPGA. For Ethernet, these transceivers are normally called the PHY chips since they provide the functionality of the physical layer. They are placed in between the FPGA and the Ethernet connector.

The way these types of communication protocols are used with FPGAs is always based on the same hardware structure:

- The medium access control layer is placed inside the FPGA by using either a specific hardwired block for most SoPC devices or a module placed in the reconfigurable fabric for all other cases.
- The physical layer is performed in the PHY chip, which requires some additional resources such as clock oscillators (dependent on the chip and Ethernet standard to be used) or passive components.
- The Ethernet connector to provide the attachment to the transmission media and galvanic isolation is placed close to the PHY device.

The connection between the FPGA and the PHY is well standardized, but there are several options depending on the chip and the transmission standard to follow—mostly dependent on the maximum transmission speed. Some of these standards are media-independent interface (MII); reduced MII (RMII), with smaller connectivity and reduced performance (up to 100 Mbps); GMII (for gigabit standards); or variations of them. Care must be taken to choose the right standard since not all media access blocks in the FPGA and all PHY chips support all of them. PHY devices also contain some low-speed control interface (such as SPI) for accessing configuration and status registers in the device from the FPGA.

Support of software-based communication layers in the FPGA is also maintained by manufacturers, providing versions of TCP/IP or UDP/IP stacks adapted for the internal soft or hard processors and OSs or bare-metal applications.

It is advised to follow some of the numerous tutorials provided as examples that accompany the SoPC tools in order to fully understand and properly design all elements in the communication stack. There are many example tools that provide simple web servers, or similar applications, in FPGA-based SoPC designs, so the time used to implement the final customized solutions is reduced. For a discussion and a practical and more detailed explanation on how fast serial I/O works on Xilinx devices, see the book by Athavale and Christensen (2005).

7.3 In-Chip Communications

This section deals with structured ways of communicating internal modules within an FPGA. Three types of connections are considered: point to point, bus based, and NoC. In this order, they are sorted from less to more complex, but also from the least to the most scalable. Lee et al. (2007) provide a quantitative analysis on the use of these three alternatives, when and how to use them, showing a theoretical basis as well as a use case example of a multimedia system.

7.3.1 Point-to-Point Connections

For a relatively small number of blocks and connections among them, point-to-point solutions are possibly the best choice. There is no connection sharing, so every pair of modules is always ready to communicate. This not only gives high communication throughput but also ensures predictable behavior from the system. Even though this is the simplest solution for internal communications, there are, however, some problems associated with it. One of them is the lack of standardization. There are less communication standards than with the other solutions presented later. However, it may run the simplest transaction protocols, as simple as "one piece of data after another," the basis for streaming, being able to very efficiently run either synchronous or asynchronous transactions. In many cases, especially for asynchronous trans-actions, FIFO memories must be placed in between both communicating sides. This FIFO memory can be made with a double-clock, double-port scheme to provide true asynchronous access between two regions with different clocks. Sufficient FIFO dimensioning ensures work independence between modules.

An important computing scheme that is feasible to be implemented with such P2P communications is the dataflow model of computation. In this model, all computing elements in the design, also called actors, follow an autonomous trigger rule such that the model starts computing whenever all incoming data from one or more connections between actors are available. Results are pro-duced when finished and sent through output P2P connections. This scheme is self-organized since there is no need of central control. Connections do not require addresses, and in some cases, data "tagging" is possible in case one actor receives different types of data elements from the same connection.

There are promising solutions based on dataflow models of computation since the lack of centralized control and predictability offer good characteris-tics for some applications. However, the main disadvantage of this solution is the increase in the communication area, so this solution does not scale well. So, for solutions that require higher and more varied connectivity, bus solutions, discussed next, or NoC solutions, described in Section 7.3.3, are better choices.

Due to the increase in the use of P2P connections with streamed data in FPGA-based SoCs, standardized efforts from FPGA manufacturers are being made. So, Avalon Stream, in the case of Altera, or AXI Stream, in the case of Xilinx, are available as connectivity solution between cores in SoPC designs. See Sections 3.5.1 and 3.5.2 for details on streaming protocols from AXI and Avalon, respectively.

7.3.2 Bus-Based Connections

In Section 3.5, specific interconnect buses for embedded soft or hard proces-sors were described in order to complement the characteristics of the proces-sors or multiprocessors themselves, embedded or embeddable, into many FPGA devices. Thus, it described in some detail different alternatives such as AMBA, AXI, Avalon, CoreConnect, and Wishbone. In this section, we are offering a different and more general view, with the purpose of showing buses together with other design alternatives that would let designers com-plement the decision criteria to choose the right solutions for their specific needs, as well as understanding in greater detail how these solutions operate and when they are required.

A bus is, in general, a bundle of wires that interconnect, in a standardized way, at least two (but normally more) elements using a shared communica-tion infrastructure. If there were just two elements connected, a point-to-point connection would normally be preferred.

Resource sharing has the obvious advantage of reducing resource usage with respect to a dense point-to-point-based complex interconnect, but has the disadvantage of reducing overall throughput and not allowing to customize and adapt bus width and performance to interconnect elements with different speed requirements (we will see later that this can be alleviated by bus bridging).

A bus interconnection involves two things: hardware aspects and a logic protocol. Hardware aspects determine the physical interconnect layer, electrical aspects, and topology of the interconnection. The logic protocol deals with timing issues, type of transactions supported, and arbitration policy, including priority schemes for multimaster-based buses. A bus may contain one or more master elements and one or more slave elements. Master elements start transactions, issue the access type (read or write), make the transaction request, and provide the address. Slave elements respond to incoming requests from a master and receive data (in the case of a write transaction) or provide data (in the case of a read) the address specified by the active master. If more than one master produces a request, an arbitration scheme must be provided in order to decide which master gets the right to make its transaction.

At the physical level, the hardware elements that access the shared medium are defined. Two basic schemes are possible, either using tristate buffers (see Figure 7.2a) or by means of multiplexed access schemes (see Figure 7.2b, which shows a partially connected multiple access scheme).

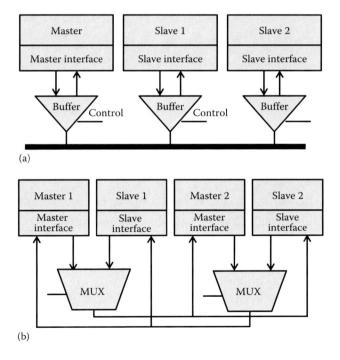

FIGURE 7.2
Physical interfaces for a bus connection: (a) tristate buffers; (b) multiplexed.

Tristate schemes are not advised because of the danger of producing contention on the bus (two elements driving the bus at the same time), which might either jeopardize the circuit or produce—for transient contentions—high current peaks. On the other side, multiplexed accesses do not have such problems, but they scale worse, requiring modifications in the structure when new elements are added or removed. Modern FPGAs, in many cases, do not include internal tristate logic (only in the chip I/Os), so multiplexed access is mandated.

The physical specification of the bus also includes the bus topology. The simplest one is the single-shared bus. They are inspired on rack-based systems, and only one transaction is possible at the same time. They may be multimaster and allow complex transactions (they will be described later), but their main drawback is the loss of performance with the increased length of the connection and the presence of multiple elements. As a rule of thumb, more than 10 different elements on the same bus will probably show some saturation and force one to use other solutions. Figure 7.3a shows an example of a shared single bus.

Other possible solutions at the physical level include crossbar switch or ring-based topologies, shown, respectively, in Figure 7.3b and c. They overcome the problem of added loss of performance, but the first one significantly increases resource utilization by the provision of multiple paths between masters and slaves, and the second one, while reduced in resource usage, produces added high latencies.

When the number of interconnected elements is high, or they can be sorted into different speed requirements and degrees of utilization, bridged and hierarchical structures offer better results. Figure 7.4 shows a bridge-based topology formed by two buses.

Bridges, like in Figure 7.4, have the advantage of providing parallel access in every segment bus, increasing overall bandwidth. However, bridging may also be used to group components that operate at different speeds into several buses, such that the design in every bus is tailored to the speed required. For instance, complex access schemes on wide buses may be used for high-speed components, while lightweight buses can be used for low-speed peripherals, which only access registers and do not provide DMA features for complex transactions. Figure 7.5 shows a three-level hierarchical bus scheme, linked by two bridges, with highest access rates at the upper level and lowest rates at the bottom. Bridges work as slaves on the upper side and as masters on the bottom side, according to the figure.

Apart from the physical level, buses also standardize the logic protocol to perform the required transactions. Timing and arbitration are the main elements to be defined. Regardless of what timing is to be used in the bus—relationships between signals to ensure correct operation—all buses fall into two different categories: synchronous or asynchronous. In synchronous buses, all timings are referred to a master clock signal, which is required to reach all elements in the bus. This issue is, at the same time,

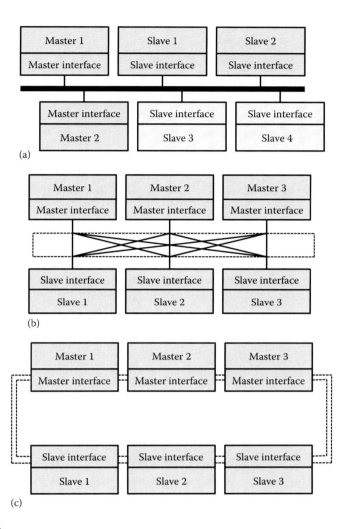

FIGURE 7.3
Bus topologies: (a) single-shared bus; (b) crossbar switch; (c) ring-based bus.

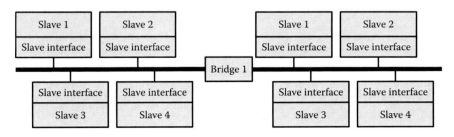

FIGURE 7.4
Bridged bus topology.

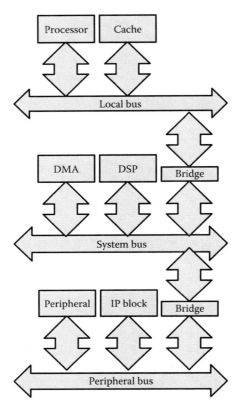

FIGURE 7.5
Hierarchical three-level bus example.

its main drawback since synchronization mismatches and skew problems may appear in long high-speed buses. They are, however, much simpler and provide faster access times than asynchronous buses. On the other hand, asynchronous buses do not have a clock, and control is effected by events in specific control signals. This ensures greater compatibility with a wider set of peripherals and modules in general, providing better timing adaptation. As a drawback, the control is more complex since it needs to set handshaking protocols between masters and slaves.

Transaction protocols also define the sequence of functional operations that need to be followed in the transaction, no matter whether it is a simple or a complex (burst) one. This is of particular importance when dealing with multimaster buses. In essence, if one or more master modules want to start a transaction, they do an "arbitration request" (AR in Figure 7.6). The arbiter solves the contention in the ARB cycle, deciding which is the master that will be given the access next, according to the priority or arbitration policy. The master that is granted the access makes a request—RQ in the figure—and the

Clock Cycle	Operation	Bus
1	AR	Free
2	ARB	Free
3	AG	Free
4	RQ	Busy
5	Busy	Free
6	Busy	Free
7	ACK	Busy

FIGURE 7.6
Single-bus transaction protocol.

slave may be busy during some cycles until data are ready, which is notified by asserting an acknowledgment signal, which ends the transaction.

These stages can be overlapped between different transactions in order to maximize bus utilization. In this case, pipelined bus structures—more complex but efficient—are required. Figure 7.7 shows an example of a multiple access in a pipelined bus structure. As can be observed, bus utilization is increased significantly, although the arbitration and grant process becomes more complex.

The number of cycles of a granted transaction may be fixed or variable. Arbitration may be centralized or distributed, and different arbitration policies may be set. These policies may be random, with static priorities, or based on periodic priority assignment, such as round robin. While this technique is suitable for distributed arbitration, which provides better scalability, it has the disadvantage of producing potentially large latencies, which makes it unsuitable for critical systems, where static priorities are preferred.

Clock Cycle	Transaction 1	Transaction 2	Transaction 3	Bus
1	AR			Free
2	ARB	AR		Free
3	AG	ARB	AR	Free
4	RQ	STALL	ARB	Transaction 1
5	Busy	AG	AG	Free
6	Busy	RQ	STALL	Transaction 2
7	ACK	Busy	AG	Transaction 1
8		Busy	RQ	Transaction 3
9		ACK	Busy	Transaction 2
10			Busy	Free
11			ACK	Transaction 3

FIGURE 7.7
Single pipelined bus transaction protocol, with three overlapped accesses.

7.3.3 Networks on Chip

The huge overhead produced in P2P communication for a sufficiently large number of interconnected devices and the performance degradation of bus-based structures forced a new paradigm for large systems with multiple interconnection needs: NoC. The main advantage with respect to their predecessors is that there is no speed degradation with size since all connections can be made for local and short distance, while being able to achieve parallelism, given the fact that several nodes can communicate in parallel by using multiple paths. It is clear that not all nodes will be able to communicate at the same time due to resource sharing, but communication policies under some circumstances may be used to ensure, in a predictable manner, sufficient bandwidth for all possible communications between nodes within the NoC.

An NoC consists of a series of links, connected by routers that interconnect multiple cores. Links are normally pairs of bidirectional channels connecting a pair of routers. The access of cores to the NoC is done on the routers, so, whenever an NoC topology drawing is observed, it must be considered that that router has an extra input/output pair of channels for the core. For instance, Figure 7.8 depicts the two most commonly used NoC topologies—a mesh and a torus, which, in essence, is a mesh connected with the same scheme of adjacency as in a Karnaugh's map. In the figure, routers in the middle of the mesh have five links (or ports), four for the visible cores, plus one for the core attached to it. All routers in the torus topology have five link connections.

Apart from these topologies, there are other topologies such as tree, fat tree, ring, octagon, and spider. However, the design and features of an NoC are not solely dependent on the topology. There are many other factors, related to the way packets are formed, how link handshaking is achieved, and how

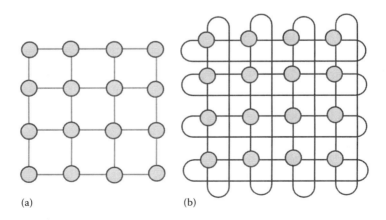

(a)   (b)

FIGURE 7.8
(a) Mesh NoC topology and (b) torus topology (a mesh variation).

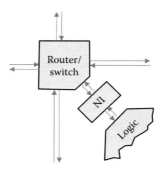

FIGURE 7.9
Access to/from logic to node router/switch.

packets are routed—including how they are switched and buffered—as well as regarding control flow and arbitration policies, which are even more important than the topology itself. For each of these factors, many different techniques have been proposed and verified, up to the point that a detailed description of all possible individual techniques, as well as a description of the collection of existing commercial and academic NoCs, is out of the scope of the contents of this book. However, some descriptions about these techniques and the problems they try to solve are mentioned later on.

Packets are injected into the NoC or withdrawn from it by means of the network interface of every core, as shown in Figure 7.9. The network interface of every core is in charge of producing—or retrieving—data according to the rules defined for the NoC. In the most general sense, data exchanged between two cores at a given time, as a result of a computation done in the transmitting node, are called messages. Messages are split into packets that, at the same time, are split into one or more *flits*, composed of several *phits*. A flit is the minimum amount of data that may be exchanged between two network elements (two routers or an NI and its router). A phit is the amount of bits that can be exchanged at a time, and it is dependent on the characteristics of the link. In contrast to conventional networks, links may be composed of a set of parallel wires transmitting more than one bit at a time.

These characteristics of the NoC are architecture dependent, except for the router internal organization, which mostly determines the main NoC communication mechanisms from a protocol perspective, not from an architectural perspective. These mechanisms are basically as follows:

- *Flow control*: It determines the node-to-node control rules and is in charge of allocating channels and buffers inside the routers to store packets.
- *Routing and switching*: Routing refers to the determination of the path between source and destination in the NoC, while switching decides how and when to connect an input and an output port within a router.

- *Buffering and arbitration*: It is related to the policies that decide which message is to be stored inside the router, either in input buffers or in output buffers, and wait for a future chance to go through an output port. Arbitration deals with the way the routers select which message has the right to go to an output port.

There are three basic types of flow control. The simplest is a handshake between two neighboring elements so that, whenever there is room in the receiver, it accepts the transmission of a packet. A more sophisticated technique, based on the earlier one, is a credit-based flow control, which relies on counting the number of packets that are sent, up to a maximum number, and reduces this number as soon as packets leave the receiving router. The third method, and possibly one of the most used ones nowadays, is the setting of virtual channels (VCs). With regard to VCs, every physical channel is shared by several logical channels, and either equal-time multiplexing or priority-based multiplexing is used as an arbitration policy to resolve which is the next packet to leave every router.

Routing protocols determine the path of messages along the NoC. Livelocks (messages returning without reaching their destination) and deadlocks (a cyclic dependency that keeps messages permanently blocked) must be avoided, yet providing some adaptability and fault tolerance—reacting to traffic conditions and permanent faults. In mesh-based or similar NoCs, routing policies are simple since traversing the network from one point to any other involves moving in any direction that gets the X–Y coordinates of the routers by which messages are passing closer to the final X–Y destination value. Dynamic routing techniques, such as "west-first, if possible," are preferred over static ones, such as "first all X, then all Y," because they offer the required adaptability.

As opposed to conventional networks, circuit switching is preferred over packet switching because it ensures predictability. The combination of arbitration policies and control flow techniques to implement VCs with priority buffering and preemption buffering is, though complex, a way to ensure predictability, and so, it is preferred in many cases over further simplified techniques. The problem with this solution is router complexity, and so, it is more suited to NoCs for integrated circuits than for FPGA-based ones, where router implementation on the fabric may consume huge amounts of resources.

Nevertheless, the complexity of the NoC design not only relies on how to select the most appropriate combination of techniques for a given problem, taking into account size, power, QoS, performance (throughput and latency), or predictability. In addition to this, NoCs must be customized in order to obtain optimal implementation for every given application. In general, if an application involves several tasks, NoCs can be adapted in shape and in resource utilization—link width adaptation and/or buffering size determination—to best fit the aforementioned requirements. Figure 7.10 shows an example of an adapted network, with four tasks that involve two or three cores each, with different link sizing and connectivity.

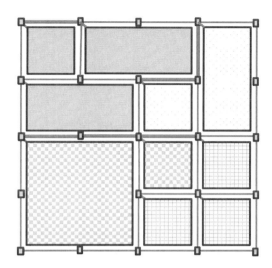

FIGURE 7.10
Customized NoC with adapted structure and connectivity.

Unfortunately, tools from FPGA vendors do not offer NoC design in their integrated environments, and among third-party tools, there are, nowadays, not so many that are purposely tailored for the FPGA design. There are some academic approaches that work with FPGA reconfigurability in order to custom design and even real-time adapt the structure and behavior of NoCs, but they are far from being a mature technology. The future might perhaps bring some NoCs designed and embedded into reconfigurable fabrics, flattening the path for the use of NoC techniques in a wider range of applications. We encourage readers whose designs have reached the point of requiring advanced communication techniques, such as NoCs, to monitor the state of development of these techniques since this will probably be one of the fields discussed in this book that has the risk of being outdated soon. Benini and De Micheli (2006) provide an excellent reference for all practical and advanced features of NoCs.

References

Athavale, A. and Christensen, C. 2015. High-speed serial I/O made simple. A designer's guide with FPGA applications. In *Xilinx Connectivity Solutions*, Product Solutions Marketing/Xilinx Worldwide Marketing Department, San Jose, CA.

Benini, L. and De Micheli, G. 2006. Networks on chip. In *The Morgan Kauffman Series in Systems on Silicon*, Elsevier.

Lee, H.G., Chang, N.C.K., Ogras, U.Y., and Marculescu, R. 2007. On-chip communication architecture exploration: A quantitative evaluation of point-to-point, bus, and network-on-chip approaches. *ACM Transactions on Design Automation of Electronic Systems*, 12:3, article 23.

8

Building Reconfigurable Systems
Using Commercial FPGAs

8.1 Introduction

The configuration possibilities offered by FPGAs created a new paradigm in digital circuit design, since the same device (i.e., the same hardware) can be adapted to provide different functions by just reconfiguring it. In other words, a device may implement different functions in the course of its operation, allowing it to be adapted to different operating conditions in response to modifications in the required functionality, changes in the environment, or even faults that might take place, therefore allowing its usability to be extended.

When hardware reconfiguration capabilities are required for a given application, the use of FPGAs offers many advantages and opportunities. Although reconfigurable systems are not limited to just FPGA-based ones, these are the most significant at commercial level. Other possibilities exist, based on custom devices with specific reconfiguration features, mainly oriented toward reconfigurable computing systems. However, these devices are intended to overcome some limitations of FPGAs in very specific areas, for instance, ultrafast reconfiguration time (i.e., reconfiguring a complete device in just one clock cycle). Coarse-grained reconfigurable architectures are an example of this, where different functions are implemented in the same silicon die so that the resulting system may be adapted to changing conditions. These solutions have limited flexibility because the functions are decided at design time and can neither be changed nor modified once the device is manufactured. On the other hand, the configurability of most FPGAs*— of truly reconfigurable devices in general—allows the functions to be performed by the system to be adapted at any moment during its lifetime, even during infield operation.

---

* Of course, we are talking about reconfigurable FPGAs. The OTP devices described in Chapter 2 also have limited flexibility.

Therefore, this chapter focuses on the use of reconfigurable FPGAs, the advantages of using their reconfiguration capabilities concurrently with normal operation (i.e., at run time), the different reconfiguration alternatives, and the existing commercial and industrial approaches, as well as the authors' view about what the future role of FPGAs in reconfigurable systems may be, through two examples, one on reconfigurable hardware acceleration and another on evolvable hardware.

8.2 Main Reconfiguration-Related Concepts

The concept of FPGA reconfiguration is, in essence, very simple: By just rewriting the contents of the configuration memory, the functionality of the device may be modified. This would provide the same flexibility as software does, but the reality is far from being that simple. Currently, there is not enough support—in terms of tools and standardization—to easily design reconfigurable systems. There is no sufficient support from the vendors to seamlessly integrate reconfiguration into the design flows either. Support is slowly being provided by academic efforts in applying reconfiguration for specific purposes, but not in a generic way.

The first FPGA devices could only be (re)configured by downloading into them a full bitstream, which would overwrite all configuration bits in the device. The configuration had to be static and performed immediately after system power-up. However, an increasing number of FPGAs, in particular those based on SRAM technology, are now allowing the so-called partial bitstreams to be downloaded, so reconfiguration can be applied just to some portions of the device (what is referred to as partial reconfiguration), even while the rest of it keeps working normally (a feature called run-time-reconfiguration, RTR). Systems having this last feature are said to be run-time reconfigurable systems (RTRSs), and those having both are called partial run-time-reconfigurable systems (PRTRSs).

Partial reconfiguration is a valuable feature for systems operating in environments where applications cannot be interrupted while the system is being reconfigured. It is also suitable for highly parallel systems that can time-share the same FPGA resources. Without this capability, it would be necessary to stop system operation during device reconfiguration and to reconfigure the entire FPGA to support a different application.

A special subset within RTRSs is composed of systems that can reconfigure themselves, which are referred to as self-reconfigurable systems (SRSs). The availability of internal configuration ports, embedded in the logic fabric, such as the Internal Configuration Access Port (ICAP) or the Processor Configuration Access Port (PCAP; managed from an embedded hard processor as in the case of Zynq devices, discussed in Section 3.3), allows self-reconfiguration to be easily carried out. Being a specialized hardware block within the silicon, PCAP has two main advantages over ICAP: It is available and ready to be used at any time (even at boot time, because it does not need to be configured in the FPGA fabric), and there is no risk of overwriting the logic that interfaces with the configuration logic.

Flash memory-based FPGAs are also capable of achieving both full and partial reconfigurations. However, care must be taken with the number of reconfigurations since flash memory contents cannot be rewritten too many times. For instance, while flash memories may usually support 10,000–100,000 programming cycles; this figure is reduced to around 1,000 times for space-qualified devices.

Applications requiring extensive reconfiguration should only be implemented in SRAM-based FPGAs. Consider a low Earth orbit satellite for marine observation, with several orbits per day, which is reconfigured for image acquisition, processing, and identification while over the sea and reconfigured again when over a reception station for data compression, encryption, and transmission. A flash-based device could not be used for more than a few months. With ASIC or nonreconfigurable technologies, both groups of functionalities would be required to be permanently implemented in the silicon. These problems can be overcome by using reconfigurable technology.

SRAM-based technologies are, however, susceptible to suffer SEUs (i.e., bit-flips) not only in the application logic but also in the configuration logic. Therefore, in some applications, ECCs, modular redundancy, and other techniques must be used in order to minimize this problem. Some of these can take advantage of reconfiguration capabilities, as discussed in Sections 8.4.1 and 8.4.3.

Granularity is another important aspect in partial reconfiguration, related to the size of the functional elements that are reconfigured, both regarding the reconfiguration needs of the application and the reconfiguration capabilities of the devices:

- Large granularity corresponds to systems where complex IP cores are replaced in the logic and, therefore, partial reconfiguration affects large portions of the FPGA. There are devices where partial reconfiguration can only be applied to a significant part of the FPGA (e.g., one half), or others that use an approach based on slots, where the FPGA is divided into several, normally identical, slots, each of which can be reconfigured separately.
- Medium granularity corresponds to reconfiguration at register level, which is typically used to modify functionality in a portion of an IP core.
- Small granularity refers to the reconfiguration of a small number of configuration bits, typically affecting the values stored in an LUT, the content of a flip-flop, or resources of similar complexity.

For large and medium grain, reconfigurable regions (RRs) must be defined such that different reconfigurable modules (RMs) may be allocated into them. If a given RM is compatible with—may be allocated into—more than one RR, then the RM is said to be relocatable. Relocation procedures are available in the reconfiguration engines to address the corresponding RM configuration information into the part of the configuration memory corresponding to the target RR in each case.

Module relocation is a clear advantage in systems based on regular slots since the same function may be allocated into different regions, providing additional flexibility, a certain degree of fault tolerance (a function may be moved from a faulty to a fault-free region), and memory footprint savings, since just one bitstream is required to support all destinations, instead of one for each possible destination.

One of the major problems associated to partial reconfiguration is how to match the required reconfiguration granularity with that actually supported by the FPGA technology being used. The smallest reconfigurable area in FPGAs depends on the manufacturer (manufacturing technology and family of devices). Some FPGA families support column reconfiguration; that is, the minimum reconfigurable unit is a column of LBs (defined in Section 2.2). The problem with this solution is that when something has to be changed even in just a single element, the entire column has to be reconfigured. More recent families support rectangular reconfiguration, where the portion being reconfigured does not necessarily have to span a whole column.

The main benefits of disruptive—non-real-time—partial reconfiguration are reduced configuration times and the possibility of silicon reuse, since the same device can be used for different tasks. In addition to these, the added benefits of using PRTRSs can be briefly summarized as follows:

- Enhanced performance and system updates because, while portions of the system are being reconfigured, the rest can remain operative. Hence, there is no loss of performance in the areas not being reconfigured, at least in principle (it may be in those interacting with parts being reconfigured). Also, system updates would in principle only affect the areas where the updated functionality is being configured.

- Hardware sharing, because in addition to the possibility offered by full reconfiguration for several applications to share the same FPGA (reducing size and cost), in this case, these applications can be executed in parallel. This characteristic is gaining importance with the increasing integration level of FPGA devices.

- Shorter reconfiguration time and lower reconfiguration energy consumption, because partial reconfiguration requires smaller (partial) configuration bitstreams than full-device reconfiguration.

In this regard, it has to be noted that techniques that accelerate the reconfiguration process at the expense of increasing instantaneous power consumption usually result in overall energy savings.

• Reduced requirements for bitstream storage resources.

RTR has also some associated problems. Even though some FPGA vendors claim they use suitable techniques so that run-time reconfiguring in their devices in an area with the same configuration it had before is a glitch-less operation, there are many restrictions derived from the atomic reconfiguration unit that can be handled. For instance, if the content of a flip-flop has to be modified, all flip-flop contents in the same column for a column-based reconfigurable FPGA have to be modified. In order to do this, system execution has to be stopped, a read-back operation has to be carried out to retrieve the contents of all flip-flops in the same column, then the desired flip-flop value has to be modified, and all flip-flops in the column have to be configured back. This does not allow real RTR to be achieved, whereas if the atomic reconfiguration unit were a single flip-flop, it would have been possible.

8.2.1 Reconfigurable Architectures

There are many reconfigurable system models, most of them relying on the use of microprocessors and reconfigurable fabric. According to Al-Hashimi (2006) and Compton and Hauck (2002), there are several possible types of coupling between both parts, as shown in Figure 8.1:

• External stand-alone, where the reconfigurable hardware is a fully independent device connected to the inputs and outputs of the microprocessor.

• Coprocessor unit or attached processor unit, where the reconfigurable hardware is closer to the microprocessor than in the previous case.

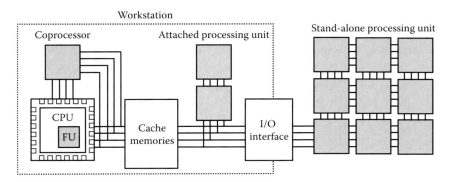

FIGURE 8.1
Different integration levels of reconfigurable logic within a processor system.

In the coprocessor approach, the reconfigurable hardware can operate as a functional resource of the microprocessor itself. In the attached processor solution, the reconfigurable hardware is accessed after the cache memories, that is, in the secondary bus.

- Reconfigurable functional unit (FU), where the reconfigurable hardware is embedded into the microprocessor. This structure is the one that most easily allows custom instructions to be added to the processor.

- Processor embedded in the reconfigurable hardware. In this case, the processor may be either soft (analyzed in Section 3.2) or hard (analyzed in Section 3.3).

8.3 FPGAs as Reconfigurable Elements

The design of an FPGA-based PRTRS involves several issues that need to be analyzed:

- Selection of a device supporting the target reconfiguration techniques. Related to this, the partial reconfiguration possibilities offered by some commercial FPGAs are described in Section 8.3.1.

- Logic partition of the device into fixed and reconfigurable areas so that reconfigurable cores, stored as partial bitstreams, can safely be allocated into suitable areas. The definition of this reconfiguration architecture is important because (as stated in Section 8.2) the process of matching the size of the reconfigurable areas with the target reconfiguration granularity requires good knowledge of the internal FPGA architecture. Hardware partitioning issues are analyzed in Section 8.3.2.

- Scalable architectures are a particular case where the resources may grow or shrink in order to adapt functionality and/or performance to changing requirements. They are addressed in Section 8.3.3.

- Partial reconfiguration requires, especially at run time, tool support for reconfiguring, adding, removing, or relocating pieces of ·hardware into different areas of the fabric. In some cases, tools have to run in embedded devices that autonomously handle their own reconfiguration. Tools supporting partial reconfiguration are described in Section 8.3.4.

- Communications between the microprocessor and the reconfigurable element, or among several reconfigurable elements if there are more than one, could be a bottleneck. In addition, the coupling between all software and hardware tasks must fulfill specific communication requirements. Choosing a suitable communication scheme is a challenge for an ASIC approach, but it is even harder in reconfigurable environments, since communication requirements may be unknown until the communication infrastructure is defined. Reconfigurable communications may be a solution to solve the problem analyzed in Section 8.3.5, where special attention is paid to NoCs and, more specifically, to reconfigurable NoC approaches (described in Section 7.3.3).

8.3.1 Commercial FPGAs with Reconfiguration Support

Altera's Excalibur were the company's first devices that allowed the whole FPGA fabric to be dynamically configured from the on-chip hard processor at any moment, by retrieving the corresponding bitstream from an external nonvolatile memory. Later, some Altera devices started to provide limited partial reconfiguration capabilities by allowing specific elements, such as serializers/deserializers or PLLs, to be reconfigured. More recently, Altera V devices (Stratix V, Arria V, and Cyclone V families) extended the support for partial reconfiguration.

A different approach is used by Atmel's FPGAs, which implement PRTR through cache logic designs, where part of the FPGA fabric can be reconfigured without loss of register data, while the remainder of the fabric continues to operate without disruption. The main drawback of these—in addition, small—FPGAs in this context is that the reconfiguration access method is bit based, which requires very low-level reconfiguration control, although it has the advantage of providing very high flexibility.

Most Xilinx SRAM-based FPGAs can be partially reconfigured. This is the reason why they are used in the majority of applications where this feature is required. Their configuration bitstream format allows a designer to modify one or more configuration packets and perform partial reconfiguration by accessing specific portions of the FPGA configuration memory. Each Xilinx device family has different reconfiguration features:

- The low-cost Spartan 3 series supports the reconfiguration of entire columns, including top and bottom I/O blocks. The first Spartan 3 family does not include an ICAP, and thus it is not well suited to designing SRSs.
- All Xilinx high-performance FPGA families provide glitch-less reconfiguration and include an ICAP. Virtex-II and Virtex-II Pro families implement column-based reconfiguration, whereas in the more recent families (Virtex-4, Virtex-5, and all Series 7 families: Artix, Kintex, Virtex, Zynq, and UltraScale), reconfiguration frames do not span entire columns, but several rows are associated with clock domains that are horizontally laid across the FPGA layout.

As for clock domains, frames for different families are of different sizes (16 rows for Spartan-6, 20 for Virtex-5, 40 for Virtex-6, 50 for Zynq and former series 7 devices, and up to 60 for UltraScale).

- Some devices have double ICAP support, which may be useful for increased fault tolerance. Zynq devices have also a PCAP, controlled from the processing system, in addition to the conventional ICAP.

Device improvements in this area are slow, mostly pushed by the research community's efforts in terms of architectures, tools, and applications.

8.3.2 Setting Up an Architecture for Partial Reconfiguration

The selection of a suitable reconfigurable device is conditioned not only by the aforementioned reconfiguration features and restrictions but also by the internal architecture of the FPGA, which has to be analyzed from its partition into three parts, namely, fixed and reconfigurable areas, described in the following, and communication infrastructure, separately analyzed in Section 8.3.5 because of its particular importance:

- The fixed area of the FPGA is the portion of the logic that does not change in any configuration. It is normally devoted to external off-chip communications, internal communication management, and self-reconfiguration. It is typically placed in FPGA regions whose irregularities prevent them from being mapped as reconfigurable areas. For column-based FPGAs, these blocks are placed in the left-most or rightmost sides of the FPGAs, and only I/O blocks close to them are used for off-chip interconnects.

- The reconfigurable area must have a fixed position because of its connections to the fixed area, but the logic inside it can be freely reconfigured. Several architectures have been proposed with different numbers of reconfigurable areas with different sizes, but for most column-based reconfigurable FPGAs, column-based reconfigurable areas are defined.

Some approaches define just one fixed area and one reconfigurable area, with different sizes and geometries. There are also some slot-based approaches, where the reconfigurable area is divided into equally sized portions of logic, with the possibility for individual IP cores to be configured in each slot. In many of these approaches, slots are column based, most of them following a 1D organization, although others with a 2D organization also exist. Figure 8.2 shows two examples of 1D and 2D partitions. A more generalized approach, based on slotless reconfigurable areas of different shapes, is also possible, at the expense of the possibility for relocating RMs in different RRs being lost. However, since RRs may have dedicated interfaces for mutually

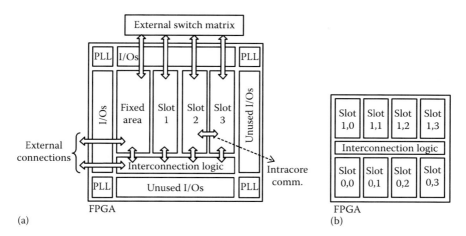

FIGURE 8.2
(a) 1D and (b) 2D architecture examples.

exclusive functions, this approach can be considered as a replacement of coarse-grained reconfigurable architectures.

An effective slot-based partition should take into account the possibility of relocating an IP core into any slot position, being capable of handling cores spanning through more than one slot, while still achieving RTR, if required.

It is common that FPGAs have some nonregular regions dedicated to memory blocks, DSP blocks, embedded hard microprocessor cores, etc. In this case, the slot partition is not trivial, and the compromise between reconfiguration granularity and number of slots has to be solved.

Some examples of proposals from the research community in this area are listed as follows:

- A platform for PRTRSs oriented to rapid prototyping of telecommunication routers and firewalls, called field-programmable port eXtender (FPX), was proposed by Horta et al. (2002). It uses a two-slot partition, which is connected through a ring network.

- The term slot for general 1D partitions was first used by Ullmann et al. (2004). A 1D multislot architecture was proposed, which uses a 1-bit serial bus for communications in the first versions (Palma et al. 2002) and a NoC in the most recent ones (Moller et al. 2006).

- Another 1D approach is described by Walder and Platzner (2004), where partial reconfiguration is performed in a fully transparent way by using a so-called hardware OS.

- The influence on hardware reconfiguration as an OS task has also been addressed by Becker et al. (2007), where hardware–software multitasking needs are analyzed.

- In the Erlangen slot machine (Bobda et al. 2005), local and shared memory accesses and off-chip interconnections are solved by using top and bottom I/O blocks of a 1D column-based partition approach.
- More recent developments follow 2D partition approaches, such as the DRNoC approach (Krasteva et al. 2008), where a method to implement partial reconfigurable partitions is presented and the communication infrastructure is a reconfigurable NoC.

In addition to these architectures, which set up the basis for reconfiguration, more versatile ones have more recently been developed. Among them, scalable architectures, described in the next section, the adaptive multithread execution platform presented in Section 8.4.2, and the evolvable hardware platform described in Section 8.4.3 are examples that deal with different connectivity or reconfiguration grain.

8.3.3 Scalable Architectures

Scalable systems are a particular case of architecture where resources can be literally added or removed from the FPGA fabric in order to adapt it to changing requirements, such as changes in execution performance, changes in the functional complexity of the problem to be solved, or adaptation to different energy budgets.

Scalability, in general, has always been pursued in digital electronic designs. Designs that can be parameterized to be adapted to changing requirements have existed for many years. For instance, any basic digital electronics course addresses issues such as how larger decoders or MUXs can be obtained from smaller ones. With "conventional" logic, these solutions can only be applied at design time since logic cannot be later modified. With reconfigurable systems, this parameterization becomes possible infield, and some applications may benefit from it.

In the context of reconfigurable computing, the construction of 1D or 2D scalable structures has the advantage of exploiting the following properties: modularity, regularity, spatial locality, and parallelism:

- Modularity, that is, the partition of a design into smaller pieces, simplifies the design process of the individual components of the ·architecture. It also reduces reconfiguration times, since only the blocks being replaced need to be reconfigured, so the overall reconfiguration times for scaling up or down a design are shorter.
- Regularity is the property that enables real scalability. If the architecture allows for module relocation, regularity results in a decrease in the amount of memory needed to store partial bitstreams, since just a reduced set of them is required to build a more complex design layout. Module relocation is of special importance for such property,

and so, regular structures in the reconfigurable fabric are desired whenever possible to increase regularity.

- Spatial locality must be enforced in order to obtain true scalability. Variable-sized structures should not require variable connectivity and connections between nonneighboring modules should be avoided. Only global signals such as clock and initialization signals can be an exception. More importantly, the connection between the "static" part and the scalable architecture should be implemented in fixed positions and with fixed, predefined interfaces. An advantage of spatial locality is that the system has more chances to achieve high operating speed since short local distances are likely to result in shorter propagation delays compared with nonregular structures with no specific locality.

- Planning execution in the scalable architecture to exploit parallelism in regular structures is also a very desirable feature. Pipelined, systolic, or similar structures are computational schemes that may allow the performance of the architecture to be maintained when scaling up the design.

Consider, for instance, an FIR filter with a direct mapping architecture for achieving high throughput. A pipelined variation of such structure, where the design is split into slices, may be obtained by inserting additional registers into the direct paths that propagate the signals. Every slice, consisting of a MAC unit and an internal register, fulfills all of the earlier-mentioned requirements for modularity, regularity, spatial locality, and parallelism, and thus, it becomes a suitable solution for a scalable reconfigurable architecture. As a consequence, it can provide excellent results in terms of throughput and performance.

While this example may be considered low- to medium-grain reconfiguration, the layout in Figure 8.3 shows a larger-grain reconfigurable scalable implementation of the deblocking filter stage of an H.264/AVC/SVC decoder, proposed by Cervero et al. (2016). This stage corresponds to one of the most computational-intensive tasks in the overall video decoding process, and as such, it is worth being implemented in hardware. When doing so, it has to be taken into account that the target video coding standard includes many types of scalability—such as temporal or spatial—and it may be designed for different resolutions, image sizes, and frame rates.

The architecture in the figure follows a 2D structure, where each processing element (PE) is in charge of filtering the edges of a minimum coded unit (MCU; each of the pixel squares all transformations—such as decompression or deblocking itself—are applied to). Deblocking effects affect the edges of every MCU. They are minimized by filtering every edge-surrounding border with different strengths, which may affect one, two, or three rows and columns of pixels around every MCU. The process requires bidirectional filtering—vertical and horizontal filters—to be performed, with data

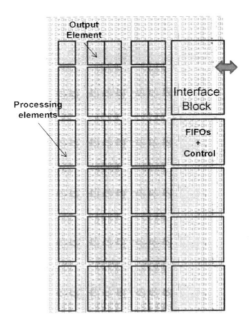

FIGURE 8.3
Architecture of a scalable H.264/AVC/SVC deblocking filter.

dependencies among the different directions. The architecture follows a systolic approach, where MCUs are computed in diagonal wavefronts and the computations are applied to horizontal image strips related to the size of the scalable architecture. The resulting performance depends on the number of PEs involved, with good linearity between the two factors until the I/O element that links the module with the static part is saturated. Experimental results showed that a variable number of PEs may satisfy the requirements from simple images, such as QCIF resolution, to 4K (ultra-HD) ones at up to 60 fps (subject to the availability of enough PEs in all cases).

Another example of a scalable architecture featuring an evolvable hardware system based on a variable-sized systolic array for image processing applications is described in Section 8.4.3.

8.3.4 Tool Support for Partial Reconfiguration

The complexity of partially reconfigurable systems requires the support from tools to automate several designs and infield operation tasks. They can be classified into the following categories:

- Tools to support design flows for generating partial reconfiguration bitstreams. There are several commercially available solutions integrated in the vendors' proprietary tools. However, these approaches are not flexible enough, in the sense that they do not handle repetitive tasks in a friendly manner, they do not perform any kind of bitstream manipulation, and deep partial reconfiguration knowledge is required of users. Also, these tools do not help in reconfigurable system simulation and debug.

- Tools to manipulate partial bitstreams so that a core can be placed into any slot in the FPGA. Partial bitstream manipulation for core relocation is a need for multislot-based architecture partitions. These tools read a bitstream that corresponds to a core placed in a specific slot position and produce another bitstream for another slot position. There are many tools of this kind derived from the JBits application (Guccione and Levi 1998), but since the application is Java based, it is difficult to have these tools running on feature-restricted devices such as embedded reconfigurable systems. However, the possibility of fine-grain reconfiguration by using low-level reconfiguration functions—mainly for LUT modification and wire rerouting—produces very complete and good results. Other tools that can run without underlying JBits support may be executed with low CPU cost on restricted embedded processors.

- Toolsets, which may also be embedded and packed as hardware-aware OSs, to support RTR. Available commercial solutions seldom cover the most basic tasks of programming and reading back configuration files. Therefore, several academic solutions have been adopted, from simple control systems implemented either in hardware or in software running in the embedded processor to complete hardware OS-based solutions. Furthermore, some solutions extend already existing OSs, such as Linux.

Some recent approaches combine specific architectures and methods to support partial reconfiguration in an efficient way:

- ReCoBus-Builder (Koch et al. 2008) is a tool chain that automates the design of systems supporting dynamic partial reconfiguration (DPR), focused on the implementation of communication infrastructures compatible with the run-time integration of partially reconfigurable modules. The supported architectures follow 1D or 2D models, allowing modules of different sizes to be stacked by attaching RMs in contiguous reconfigurable elements that communicate through either buses or point-to-point connections.

- OpenPR flow (Sohanghpurwala et al. 2011) is an alternative to the partial reconfiguration flow from Xilinx, offering a similar functionality. It is based on Torc (Steiner et al. 2011), which offers an API to manage logic netlists in standard EDIF format and physical details of these netlists in XDL or NCD formats (both from Xilinx) and allows bitstreams to be manipulated. By using Torc, OpenPR automates the design of DPR systems by means of a set of scripts that process user-provided XML input files. The tool implements constrained placement and then uses a technique called "blocker macros" to guide routing (placement and routing are described in Section 6.2.3.3). After routing and full bitstream generation, a partial bitstream for the reconfigurable module instantiated in a reconfigurable area is obtained. Researchers from the same group proposed an alternative to OpenPR, known as "Wires on Demand" (Athanas et al. 2007), which supports reconfigurable run-time intercommunications.

• GoAhead (Beckhoff et al. 2012) is an evolution of ReCoBus-Builder, with a graphic interface and area estimation tools that facilitate floor-planning. Routing is solved by using the aforementioned blocker macros and relies on XDL for design support at physical level. Apart from bus macros, direct wiring is supported, which allows fine-grain reconfiguration to be exploited with no increased area over-head. Wires crossing a border between two regions (reconfigurable or static) are set to use specific wires in such a way that different RMs exactly match the desired connectivity, with no additional resource utilization. The design of the static system and the RMs are independent processes in GoAhead. Module relocation is also compatible with this flow.

8.3.5 On-Chip Communications for Reconfigurable System Support

On-chip communications are an important challenge for all SoC designs, but this problem is even more important for reconfigurable systems since the communication needs may change, and may even be unknown, for future hardware configurations. Therefore, scalable and flexible communication structures are needed in this context.

Hardware tasks need to exchange data between them, as well as with off-chip components. Usually, the fixed area of the internal FPGA architecture partition is used for this last purpose. The communication infrastructure is in charge of linking slots between them, as well as slots with the fixed area. As for the ASIC intracommunication problem, buses and NoCs are the most frequently implemented solutions.

Hardware tasks, wherever they are placed, have to be designed with fixed-position connections. Many alternatives have been proposed in order to increase the connectivity of modules. They allow either bus or point-to-point connections to be used, providing access to a NoC infrastructure or real point-to-point links.

The move from bus to NoC approaches has been followed also in the reconfigurable system area. The number of slots for past FPGA technologies was not high enough to justify, in most cases, the need for a NoC because buses are a simple and flexible solution for connecting a low number of cores. However, as FPGA capacity increases, so does the number of cores that can be implemented in a single device, and NoCs are a promising solution for these larger systems.

NoCs in reconfigurable systems are typically associated with 2D FPGA partitions (described in Section 8.3.2), with both regular meshes and heterogeneous networks being considered. Work is being conducted to verify the possibility of reconfiguring not only the cores themselves but also the communication infrastructure so that the available communication resources can be made to fit the variable communication needs for just one given configuration or for a different set of them.

The solution by Ullmann et al. (2004) has switch matrices that can be reconfigured. The latest works on the Erlangen machine (Bobda et al. 2005) show the use of a reconfigurable NoC, called DyNoC, which allows core grouping and the network reconfiguration to bypass the portions of the NoC that are used by the merging of two adjacent cores. The CoNoChi (Pionteck et al. 2006) and DRNoC (Krasteva et al. 2008) NoCs are more flexible solutions since network interfaces and some parts of the routers can be modified. DRNoC may reconfigure switch matrices, network interfaces, and routers' parameters, enabling not only NoC communications but also a combination of these with point-to-point connections and bus-based solutions.

8.4 RTR Support

Reconfiguration at run time requires some elements to decide what, when, and where reconfiguration is to be applied in order to add, remove, or replace a given module with a different one. Run-time support has to make decisions based on optimization criteria, such as performance maximization or reduction in energy consumption, while considering restrictions such as limited resource availability or reconfiguration overhead. Since the problem is not trivial, reconfiguration is managed by a program running on a processor, either embedded (if self-adaptation is pursued) or external (using off-chip communication channels).

It can be easily observed that there are similarities between handling software tasks and hardware tasks in an OS. Both types of tasks can be loaded or offloaded according to priorities in the execution, scheduling of tasks, or resource-sharing restrictions. Therefore, research efforts have been devoted to providing reconfiguration support for hardware task management at OS level in a similar way as is provided by conventional OSs for software tasks. The added complexity of reconfiguration-aware OSs is depicted in Figure 8.4.

All OS features related to software task execution are inherited by the execution of hardware tasks, so the additional issues are related to solving

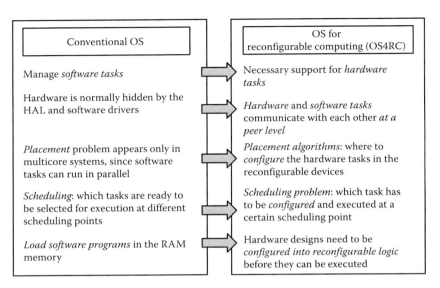

FIGURE 8.4
OS requirements for reconfiguration support.

the allocation of hardware tasks to deliver proper acceleration of multiple tasks, coming either from a multicore system or from the concurrent execution of multiple tasks in a single core. Usually, accesses to reconfigurable hardware accelerators rely on buses or similar structures, which need to be shared among all hardware tasks, with associated bottlenecks to be solved. Hardware task scheduling must also be determined since the penalty incurred when reconfiguring hardware tasks must be accounted for to determine execution efficiency and potential energy savings. Also, hardware reconfiguration has an impact on bandwidth utilization within the structure. Since partial bitstreams are prefetched from nonvolatile memory to a faster one (such as an external DDRAM) to minimize reconfiguration time, it must be taken into account that data required for task execution and bitstreams for hardware reconfiguration might share internal paths.

All these problems might be solved with custom solutions. However, it is possible to envisage layered structured approaches capable of addressing them in an efficient way. The structure in Figure 8.5 shows a six-layer approach for applying reconfigurable computing in hardware-aware OSs:

1. The application layer is in charge of communication and synchronization between tasks.
2. The module layer manages partial bitstreams, including their possible prefetching into faster intermediate memories.
3. The scheduling layer determines the best candidate(s) to be set into hardware, according to the available resources.

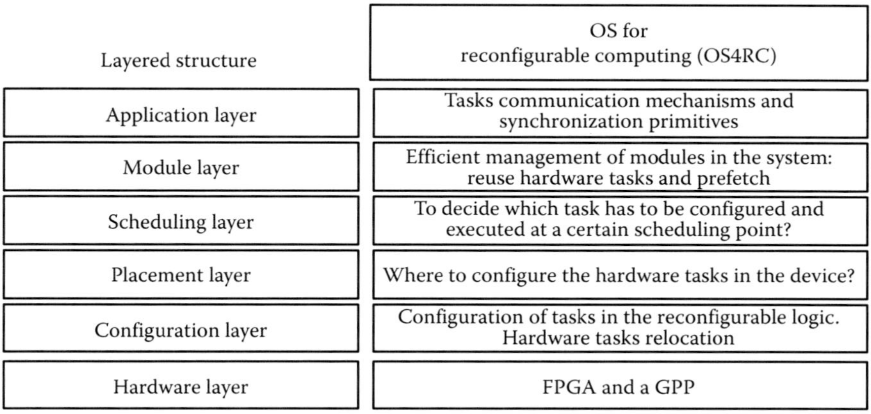

Layered structure	OS for reconfigurable computing (OS4RC)
Application layer	Tasks communication mechanisms and synchronization primitives
Module layer	Efficient management of modules in the system: reuse hardware tasks and prefetch
Scheduling layer	To decide which task has to be configured and executed at a certain scheduling point?
Placement layer	Where to configure the hardware tasks in the device?
Configuration layer	Configuration of tasks in the reconfigurable logic. Hardware tasks relocation
Hardware layer	FPGA and a GPP

FIGURE 8.5
Layered structure for RTR.

4. The placement layer determines the best possible position for a hardware task, according to intrahardware task connectivity (if available), to the presence of faults in some RRs, or just to minimize reconfiguration overhead.

5. The configuration layer includes a reconfiguration engine that reads the bitstream for a given task to be implemented in a given position or relocates a task to the desired new placement and performs the reconfiguration process.

6. The hardware layer is the element supporting reconfiguration at physical level. In the case of FPGAs, such support is provided by their reconfiguration capabilities.

How much advantage can be taken from reconfiguration capabilities is directly linked to how efficient the reconfiguration management is. External managers providing remote reconfiguration mechanisms are usually associated to low reconfiguration rates, whereas the most interesting and challenging situations appear when the reconfiguration capabilities of a system are self-managed. Self-managing systems are described in Section 8.4.1, highlighting the possibilities and added value a system may get from the addition of reconfiguration technology at different levels. In Section 8.4.2, an example is presented on how dynamic reconfiguration may impact a multithread execution-capable architecture by adapting it to different levels of execution performance, fault tolerance, or energy consumption.

8.4.1 Self-Managing Systems

Self-management refers to the ability of systems to manage themselves according to high-level objectives, reducing external intervention. This term is equivalent to *self-adaptable* or *autonomic computing*, according to the terminology coined by Paul Horn (Kephart and Chess 2003). According to the *Oxford Dictionary*, adaptation in biology is *the process of change by which an organism or species becomes better suited to its environment*. In the same way, generalizing the definition for self-adaptive software by Salehie and Tahvildari (2009), a self-managing system can be defined as one that adjusts various artifacts or attributes in response to changes in itself or in its context.

In the end, the goal of self-managing systems is to reduce human intervention in the management of the system, which is a very important burden (sometimes even higher than the initial cost) in current, increasingly complex and ubiquitous computing infrastructures. Self-managing systems would allow complexity to no longer be considered as the main system limitation, as it is today. The ideal situation would reduce human intervention to the definition of high-level goals and policies. In fact, the self-managing concept includes system integration, installation, configuration, tuning, and maintenance, among other tasks. To succeed in these, a self-managing system should be capable of not only dealing with changing components, workloads, and demands but also adapting to changing external conditions, including malicious attacks against hardware or software. Following the aforementioned Horn's terminology, responses to these events are required to be not only automatic but also autonomic. Again from *Oxford Dictionary*, autonomy is *the right or condition of self-government*, which is shown by those systems that are *freed from external control or influence*. In fact, this term was selected as a reference to the autonomic nervous system. In the case of humans, this system controls some vital body parameters, such as temperature or heart rate, without expending conscious brain capabilities, which remain available to carry out emotional processes, either rational or irrational, such as decision-making.

Features required from autonomic computing systems are also called the self-* properties. Kephart and Chess (2003) discern four main properties:

1. Self-configuration, which is the seamless configuration of components and systems according to predefined rules and goals, triggered by changes in the environment. Since those changes are dynamic, so has to be the configuration. Moreover, in most cases, system services cannot be disrupted during reconfiguration. Therefore, the importance of reconfigurable hardware for autonomic systems can be ·clearly noticed.

2. Self-optimization, through continuous exploration of ways for performance and resource utilization to be optimized, in order to identify the best possible combination of system parameters. This feature mainly moves the onus from design time on to operating time, not only to evolve from static to dynamic behavior but also from reactive to proactive systems, which are capable of anticipating and foreseeing changes.

3. Self-healing, which consists in detecting, diagnosing, and repairing system disruptions and failures, in order to maximize dependability. Fault prediction and prevention tasks are included here.

4. Self-protection, understood as a set of countermeasures for the system to defend against malicious attacks, as well as against bugs not fixable by self-healing mechanisms. To the extent possible, self-protection should also work in a predictive way.

Around these fundamental ones, many other "self-*" properties have been identified, such as self-governing, self-organizing, self-diagnosis, or self-recovery. A clear requirement is therefore for systems to be capable of continuously acquiring information about themselves and their environment through self-monitoring. In other words, self-awareness and context awareness mechanisms are required.

The combination of self-management techniques with the flexibility offered by dynamic and partial reconfiguration opens up new research opportunities (Santambrogio 2009). On one hand, adding reconfigurable hardware to a self-managed system introduces an extra degree of freedom. Hence, hardware becomes a dynamic component that can be tuned to contribute to the fulfillment of the overall system specifications. On the other hand, adding self-management capabilities to a reconfigurable system allows the inherent complexity associated with the design and run-time management of these types of systems to be reduced.

Autonomic systems offering all the features described earlier are still a chimera. There is a long way ahead to achieve such a degree of autonomy. Meanwhile, Steiner and Athanas (2009) proposed a classification of systems according to their level of autonomy. Although oriented to aerospace systems, it also serves as a general roadmap to guide the evolution of this technology. Their classification is organized into the following levels:

- *Level 0*: No autonomy at all. The designer is the only person responsible for updating the system, whereas this is completely passive, without any knowledge about reconfiguration issues.

- *Level 1*: Systems have some (limited) reconfiguration-related information, including resource utilization and free area. They are even capable of creating simple connections between RMs. The 1D slot-based approach is the typical model at this level.

- *Level 2*: Systems are capable of placing and routing their own netlists that, once processed, can be configured according to an internal model. In this case, architectures are more efficient and sophisticated than slots.

- *Level 3*: Synthesis is also an autonomic system feature, and therefore, component descriptions can be behavioral, which implies a huge reduction in the associated engineering costs.

- *Level 4*: Self-awareness features are included in the system, which at this level is capable of detecting and monitoring conditions of interest, but not of responding to them.
- *Level 5*: A response library is available within the system, gathering possible responses to observed events. The library may be shared with and augmented by other systems.
- *Level 6*: Systems are capable of applying responses, by synthesizing and implementing the behavior corresponding to expected results. Hence, responses are increasingly complex.
- *Level 7*: Systems are capable of extending and adapting response libraries, by inferring required behavior from detected conditions. In other words, systems can apply some computational intelligence techniques to reconfiguration-related tasks.
- *Level 8*: Systems are capable of learning and, in this way, deciding if applied changes are satisfactory. If yes, they are introduced in the response library.

The state of the art is far from providing systems efficiently covering all these levels, and in addition, not all applications fit in one or another of them. There is a lot of room for new developments in this area, but, undoubtedly, reconfigurable devices are well suited to addressing different levels of adaptation in autonomic system design.

8.4.2 Adaptive Multithread Execution with Reconfigurable Hardware Accelerators

As discussed in Section 6.5, the need for higher performance in embedded systems is demanding the use of languages, such as CUDA or OpenCL, allowing designers to take advantage of all possible parallelism that can be identified in algorithms. Traditionally, these algorithms have been implemented in multicore systems or GPGPUs, but FPGAs are also suitable platforms for them.

The parallelism between the programming model, the memory model, and the architecture model allows the design to be implemented in a fixed number of CUs, each one consisting of a fixed number of PEs such that every single work-item/thread is executed in a dedicated PE, so a set of PEs may share local memory in the CU bundling them. CUs exchange data with external memory for memory-intensive tasks. The scalability provided by the independent execution of work-groups/thread blocks in CUs allows the kernel execution to be mapped into any arbitrary number of CUs.

Let us elaborate more on the terms "fixed number of CUs" and "fixed number of PEs per CU." A reconfigurable architecture may accommodate an arbitrary number of CUs, so it may ideally support an arbitrary number of hardware accelerators. In such a system, the execution platform of a given kernel can be chosen or modified at run time.

There are two main advantages to this approach: First, the program running in the host does not depend on the way the execution of the kernels is distributed among a variable number of CUs. Second, performance and energy consumption are a direct function of how many resources are available in the system and the execution requirements set for the different tasks at run time.

As one might guess, the higher the number of accelerators, the higher the power consumption. However, acceleration reduces computing time, so eventually energy savings may be obtained. This advantageous trend may persist up to the point memory access bandwidth is saturated. For a certain number of hardware accelerators, memory transfers will use memory interfaces at their full capacity. From that point, no further improvement will result from the use of additional accelerators, but overall consumption and execution time will increase instead.

From the reconfigurable system viewpoint, module replication is a selective way to set different operation modes in a system. For instance, if, instead of assigning different tasks to different CUs, the same tasks are assigned to two or three CUs, advantage can be taken from module replication to build dual modular redundancy (DMR) or triple modular redundancy (TMR) configurations targeting increased fault tolerance. Little modifications would be required on the way data are delivered to the CUs, with the exception of the need for a comparator or a voter, which may be included at memory transaction level (when writing results from the CUs into memory).

In summary, in a reconfigurable architecture, the combination of variable operation points, energy consumption tuning, and performance-fault tolerance trade-offs is enabled by appropriate reconfiguration management. One such architecture, called ArtiCo³, is proposed by Valverde et al. (2014). Figure 8.6 shows the main components of this architecture, which are described in the following:

- *Host processor*: Contrary to the general approach followed by CUDA or OpenCL in the context of HPC systems, which considers host and device as separate entities, in the context of embedded hardware acceleration, they may be placed together in the same device. The host is the embedded processor (single or multicore) running the application code(s) that requires kernels to be accelerated in hardware.

- *Resource manager*: It is the module in charge of finding the operation point depending on both internal and external conditions. It schedules the use of resources to work within the different operation modes supported by the architecture.

- *Data shuffler*: It is the module in charge of data transactions with the reconfigurable kernels. The way data are delivered to and taken from them depends on the operation mode defined by the resource

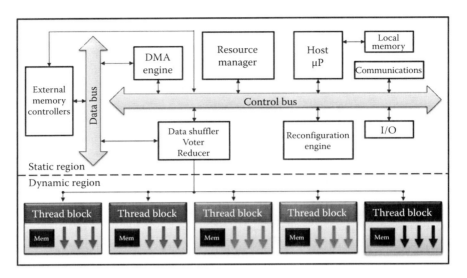

FIGURE 8.6
Dynamically reconfigurable architecture for multithread acceleration. Each CU (bottom of the figure) executes several threads within each dynamically reconfigurable hardware accelerator.

manager. Data collection is complemented by a voter that checks for discrepancies among blocks operating in either DMR or TMR configurations, generates a congruent output, and reports errors to an optional fault management module.

- *CUs*: Each work-group/thread block is mapped into a CU, which contains a wrapper module to interface with the static distributed logic, memory, and register interfaces to exchange data between the static region and/or global memory (on the static side) and a multi-bank memory (serving as local memory for the work-group/thread block) and configuration and control registers (on the CU side).

- *Reconfiguration engine*: It includes the controller of the ICAP, managed from the resource manager.

- *Interfaces with external components and memory controllers*: Memory transactions are optimized by the use of full versions of the AXI4 bus (described in Section 3.5.1.3) such that utilization of the external RAM bus as close to maximum as possible is achieved.

The resource manager is the brain of the architecture, which can be analyzed at two different levels. At the first level, operating points are determined according to three parameters: consumption, computation, and dependability needs. The actual value of each one of these parameters depends on both external and internal conditions. For instance, in a platform whose architecture is customized for wireless high-performance distributed networks, the main conditions to be evaluated are battery level, current consumption per power rail, radio messages, or number of faults detected. With three parameters being considered, the solution space is a surface within a cube whose axes represent pure solutions for fault tolerance, security, and acceleration, respectively.

At the second level, the resource manager works as a task scheduler organizing the tasks demanded by the host. To achieve an intelligent use of available resources, it has to make decisions about what kernels to invoke, the number of blocks per kernel working in parallel, the number of work-items/threads per work-group/thread block, and the amount of information to be delivered.

Figure 8.7 shows two graphs that illustrate the dynamic adaptation possibilities of this architecture (Rodriguez et al. 2015). They show energy versus execution performance and energy savings versus number of accelerators, respectively, for an AES-256 encryption algorithm with variable fault tolerance (simplex, DMR, or TMR) and different work-items/threads per work-group/thread block. Both the number of accelerators and the fault tolerance structure can be dynamically configured, whereas the number of PEs per CU has to be decided at design time, since an HLS synthesis process is required to obtain the relocatable partial bitstream with the hardware elements of a CU.

Many embedded control systems are very demanding in terms of dependability, performance, and power budget requirements. In addition, these requirements may vary over time, as implied in the satellite example in Section 8.2. Therefore, dynamic adaptation is crucial. The application code can be configuration agnostic, whereas the resource manager can coordinate with context-aware functions to determine the best configuration at each moment.

The ARTICo^3 architecture may allow a parallel kernel invocation method to be implemented, which would launch parallel work-groups/thread blocks in a variable number of accelerators as they are progressively being reconfigured, measure execution time for a work-group/thread block in just one accelerator, feed this measurement back for more precise time prediction, or decide the optimal number of blocks in order to satisfy a deadline.

8.4.3 Evolvable Hardware

Evolvable systems, as its name correctly evokes, may autonomously decide to generate new designs, in response to changes in the functional specifications, external conditions, or the system itself (for instance, the presence of faults). They reach high levels of autonomy in the classification from Steiner and Athanas (2009) included in Section 8.4.1. Evolvable systems are a type of bioinspired systems, in the sense that they imitate the evolution of species, capable of adapting generation by generation to changing conditions, increasing their survivability.

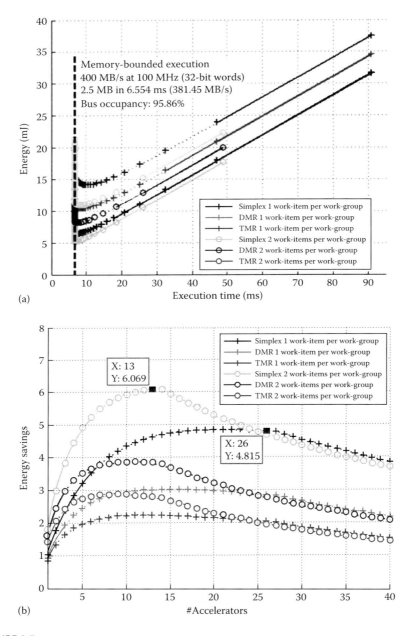

FIGURE 8.7
Space exploration graphs: (a) performance vs. energy for variable number of accelerators and redundancy mode and (b) maximum energy savings vs. number of accelerators.

The inspiration in biology is taken up to the point that circuit modifications are originated by changes in an associated "chromosome" that describes the structure and/or functionality of the circuit. Chromosome changes are produced by genetic operators, such as mutations from the parents' chromosomes or crossover of two (or more) chromosomes. Species evolve because their fittest individuals survive better. Their offspring are supposed to inherit the "good and bad" characteristics of parents, and, again, those who are better suited to survive or who perform better according to some required characteristics (for instance, running faster, being stronger, or flying longer) will iteratively produce generations better suited to their environments.

If this concept is generalized and applied to hardware, we may think in terms of an evolutionary loop that results in circuits mapped according to a "modifiable" chromosome, improved by selecting those that, after an evaluation process, are identified to be better suited to perform a given task. The evaluation process is carried out by putting the proposed circuit to work and use a "quality" function—called fitness function—to determine, in a quantifiable manner, some specific characteristics of the circuit such as performance or proximity to a "golden" solution. A "soft" metric* is required so that the evolutionary loop may gradually converge and come up with an individual (a circuit) with the target behavior. For instance, for a marathon runner, it could be the time needed to complete the race, whereas for a weightlifter, it could be the lifted weight.

The determination of a suitable fitness function is not always evident. For the former examples, the target would be to minimize time or maximize weight, respectively. But how to extrapolate this to circuits? Miller (2011) designed combinational circuits by evolutionary techniques using a fitness function that measures the number of correct outputs generated by each circuit. If we consider, for instance, a four-input combinational circuit, which has 16 different input combinations, the goal is to select circuits reaching a fitness value of 16, that is, those among all generated circuits that fully comply with the expected functionality.

This technique, however, cannot be generalized for much larger circuits because the complexity of the design and the huge space of possibilities would render the evolutionary process very slow. For large circuits, however, some custom functions have shown to be appropriate for specific designs. In these cases, the fitness functions evaluate the similarity of the output coming out from the proposed circuits with respect to the expected output. For instance, in image denoising applications, an evolved circuit might be evaluated by feeding it with a noisy image (resulting from adding predefined amounts of certain noise types to a known noise-free image) and comparing the image generated at the output with the noise-free one. Fitness functions could then be based on metrics such as peak signal-to-noise ratio or structural similarity index.

---

* Here "soft" means the metric cannot be a "yes or no" function, but one whose result indicates how far or close the circuit is from the golden behavior.

Evolution must be generalizable. This means that, after the training process, the resulting circuit must achieve its expected function for any other input different from the one used for training. Figure 8.8 illustrates different evolutions of an image filtering and processing circuit (Salvador et al. 2013). Each row shows the image the circuit was trained with, the resulting image after an evolutionary run, and the reference image for the training, as well as a different input image and the result of applying to it the functionality provided by the evolved circuit. Results show that the circuit is generalizable in that it can be adapted to different functions and its behavior is preserved for different input images:

- The first row highlights the good behavior of the circuit for salt-and-pepper (S&P) noise.
- The second row shows it also performs well for burst noise affecting several consecutive pixels.
- The third row demonstrates its adaptability to a different problem. If the reference image is set to be the edge of the input image, the circuit evolves to perform edge detection.
- Finally, the fourth row shows that the circuit can evolve to combine noise elimination and edge detection features.

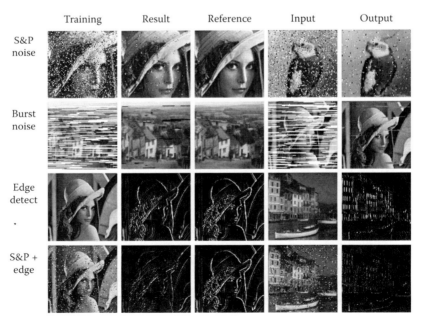

FIGURE 8.8
Adaptation and generalization of an evolvable systolic array for image processing.

The case in the third row deserves special attention. The reference image may have been obtained with any software-based edge detection algorithm (for instance, a Sobel filter). The evolution process came up with a circuit performing the same function, so it can be stated that a hardware circuit has been created, with no external intervention, from a software specification. It can be then concluded that evolution is useful not only for generating circuits but also for accelerating some parts of a system by moving them from software to hardware.

Figure 8.9 shows the architecture of the system achieving these results, which relies on the use of dynamic and partial reconfiguration. The reconfigurable portion of the architecture is a systolic array, consisting of a 4 × 4 matrix of tiny PEs, each of them capable of performing simple functions based on basic operators, such as adders or subtractors, with or without saturation, shift operations, or maximum/minimum functions. Each PE takes values from the upper or left sides (or both) and provides its results to the lower and right sides. The arbitrary composition of several PEs in the array allows the system to be adapted to different functionalities. Inputs to the systolic array are taken from a 3 × 3 pixels sliding window that moves all along the image. The pixel coming out of the array is placed in the central position of the window. The inputs of the array are taken from the moving window in an arbitrary way, decided by evolution, using a set of MUXs whose control signals are driven as specified in the circuit's chromosome. The output of the array is selected by another MUX whose control signal is also generated by evolution.

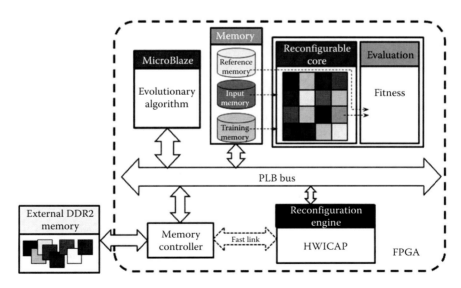

FIGURE 8.9
Components of an evolvable architecture.

The chromosome of the circuit is proposed by an evolutionary algorithm executed in a MicroBlaze soft processor. From this chromosome, a reconfiguration engine allocates the required PEs into the corresponding array positions. In order to achieve higher reconfiguration speed, PEs are available in RAM after having been copied from an external library stored in nonvolatile memory. Since PEs are relocatable, the reconfiguration engine can allocate into any array position. In this way, only one copy of each PE needs to be stored in the library.

The array is fed with an input image for training. Then, output pixels are compared with those of the reference image, and the sum of absolute errors is used as fitness function. The evolutionary algorithm fetches this value and identifies the best candidate among the offspring of the generation under evaluation (the one with the lowest fitness value), and a new candidate solution is selected. A mutation of the selected candidate's chromosome is done by changing some genes and replacing them by a random value. This is repeated for each new candidate. Mutation rates—percentage of genes to be changed—and offspring size must be adjusted to achieve good results. Diversity is key to achieving a good design space exploration, so several evolutions can be performed in parallel, not to get stuck in local minima.

There is, however, a very important issue to be considered when evolution is carried out in the same reconfigurable fabric that will be used for normal operation: If the fabric has a fault (even a permanent one), the system can evolve to circumvent it and provide improved fitness results despite its presence. By re-evolving a circuit after a fault is produced, alternative candidate solutions may be found that operate correctly in the presence of the fault. This provides the system with self-healing capabilities, which can be very important for fault-tolerant systems. Self-adaptation against faults is a natural property of such evolvable systems, referred to as intrinsically evolvable systems.

Adaptation to faults may be combined with other fault tolerance techniques in order to improve the survivability of the system. The combination of three evolvable systems in a TMR structure with a voting connected to their outputs may completely override the effect of a fault in any of the three elements, since the two nonfaulty ones will impose the result. In addition, evolution may allow the faulty module to achieve again a nonfaulty behavior, thereby minimizing the probability of fault accumulation, which could jeopardize correct functionality if faults are present in more than one module. Figure 8.10 shows the evolution of a faulty module trained by following the functionality of a fault-free one. The evolutionary algorithm has the objective of minimizing the differences between the fault-free circuit and the circuit under evolution. If, after some runs, the fitness function reaches a value of zero, it can be concluded that the formerly faulty module fully recovered from the fault and full TMR operation has therefore also been recovered.

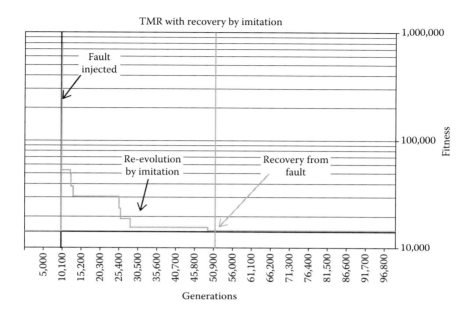

FIGURE 8.10
Self-healing of a TMR structure through "re-evolution by imitation" of the faulty module.

Gallego et al. (2013) analyzed the scalability of the systolic array discussed earlier. It is possible to scale up and down the size of the array—in one or both dimensions—in order for the system to be adapted to variable-complexity problems. Also, scalability may be used to increase resource utilization in order to overcome the effects of an increasing number of accumulated faults. An example of additional fault tolerance by means of scaling up and re-evolution is shown in Figure 8.11. After a first evolution up to a certain fitness value, a threshold value is set to determine the conditions under which scaling up the circuit is required. It can be seen that the system may, in this case, support the occurrence of three random faults without going over the threshold value. However, after the fourth fault is injected, it does not recover anymore. By scaling up one dimension, the system still does not recover, but after scaling up in the other dimension, the fitness value gets into an acceptable range; actually, it even gets better results than at the beginning since more resources are now available, allowing the occurrence of two more faults to be supported without reaching the threshold value. In brief, the figure shows that adding an extra row and column results in the system being capable of supporting up to six accumulated faults. The additional resource utilization is much better than in a conventional TMR approach. Of course, the evolutionary algorithm and the processor that runs it are a critical part of the recovery process, so they should be hardened if the system has strict fault tolerance requirements.

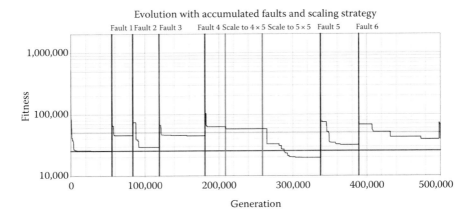

FIGURE 8.11
Increased self-healing properties by combining scalability and re-evolution.

In general, scalable reconfigurable systems aim at providing adaptability to variable-complexity problems through modular architectures that, by iteratively adding the same module, may adapt to variable performance requirements. Modularity is achieved by providing local connectivity, and in many cases (e.g., the system analyzed in Section 8.3.3), performance increases linearly with size. On top of that, local connectivity may also achieve higher operating frequencies, since local interconnections are faster than long-distance wiring.

To give some additional insight about the performance these solutions may achieve, some more data regarding this particular systolic array can be considered:

- It is capable of operating at frequencies over 450 MHz on Virtex-5 devices, a figure difficult to obtain even for regular nonreconfigurable midsized designs. This is mainly due to the fact that the PEs used are very simple, fitting in just 2 Virtex-5 LBs.

- Since systolic operation generates an output pixel in every clock cycle, the circuit may operate at over 450 Mpixels/s, with additional latency but no speed degradation if array size is increased.

- The combination of a fast reconfiguration engine with several arrays operating in parallel allows more than 80,000 evaluations/s to be achieved, as shown by Mora et al. (2015).

- All this results in systems that, after 3–5 s of evolution time, produce high-quality, high-speed, fault-tolerant image processing circuits.

References

Al-Hashimi, B.M., ed. 2006. *System-on-Chip: Next Generation Electronics.* IET Press, London, U.K.

Athanas, P., Bowen, J., Dunham, T., Patterson, C., Rice, J., Shelburne, M., Suris, J., Bucciero, M., and Graf, J. 2007. Wires on demand: Run-time communication synthesis for reconfigurable computing. In *Proceedings of the International Conference on Field Programmable Logic and Applications (FPL 2007),* August 27–29, Amsterdam, the Netherlands.

Becker, J., Donlin, A., and Hubner, M. 2007. New tool support and architectures in adaptive reconfigurable computing. In *Proceedings of the IFIP International Conference on Very Large Scale Integration 2007,* October 15–17, Atlanta, GA.

Beckhoff, C., Koch, D., and Torresen, J. 2012. Go ahead: A partial reconfiguration framework. In *Proceedings of the 2012 IEEE 20th Annual International Symposium on Field-Programmable Custom Computing Machines, (FCCM 2012),* April 29–May 1, 2012, Toronto, Ontario, Canada, pp. 37–44.

Bobda, C., Majer, M., Ahmadinia, A., Haller, T., Linarth, A., and Teich, J. 2005. The Erlangen slot machine: Increasing flexibility, in FPGA-based reconfigurable platforms. In *Proceedings of the 2005 IEEE International Conference on Field-Programmable Technology,* December 11–14, Singapore.

Cervero, T., Otero, A., López, S., de la Torre, E., Callicó, G.M., and Riesgo, T. 2016. A scalable H.264/AVC deblocking filter architecture. *Journal of Real-Time Image Processing,* 12(1):81–105.

Compton, K. and Hauck, S. 2002. Reconfigurable computing: A survey of systems and software. *ACM Computing Surveys,* 34:171–210.

Gallego, A., Mora, J., Otero, J., de la Torre, E., and Riesgo, T. 2013. A scalable evolvable hardware processing array. In *Proceedings of the 2013 International Conference on Reconfigurable Computing and FPGAs (ReConFig 2013),* December 9–11, Cancún, Mexico.

Guccione, S.A. and Levi, D. 1998. XBI: A Java-based interface to FPGA hardware. In *Proceedings of SPIE, Configurable Computing: Technology and Applications,* Vol. 3526, Boston, MA, pp. 97–102.

Horta, E.L., Lockwood, J.W., Taylor, D.E., and Parlour, D. 2002. Dynamic hardware plugins in an FPGA with partial run-time reconfiguration. In *Proceedings of the 39th Design Automation Conference,* June 10–14, New Orleans, LA.

Kephart, J.O. and Chess, D.M. 2003. The vision of autonomic computing. *Computer,* 36:41–50.

Koch, D., Beckhoff, C., and Teich, J. 2008. ReCoBus-Builder—A novel tool and technique to build statically and dynamically reconfigurable systems for FPGAs. In *Proceedings of the International Conference on Field Programmable Logic and Applications (FPL 2008),* September 8–10, Heidelberg, Germany.

Krasteva, Y.E., de la Torre, E., and Riesgo, T. 2008. Virtual architectures for partial runtime reconfigurable systems. Application to Network on Chip based SoC emulation. In *Proceedings of the 34th Annual Conference of the IEEE Industrial Electronics Society (IECON 2008),* November 10–13, Orlando, FL.

Miller, J.F. 2011. *Cartesian Genetic Programming,* Natural Computing Series 2011. Springer.

Moller, L., Soares, R., Carvalho, E., Grehs, I., Calazans, N., and Moraes, F. 2006. Infrastructure for dynamic reconfigurable systems: Choices and trade-offs. In *Proceedings of the 19th Annual Symposium on Integrated Circuits and Systems Design*, August 28 to September 1, Ouro Preto, Brazil.

Mora, J., Otero, A., de la Torre, E., and Riesgo, T. 2015. Fast and compact evolvable systolic arrays on dynamically reconfigurable FPGAs. In *Proceedings of the 10th International Symposium on Reconfigurable Communication-Centric Systems-on-Chip (ReCoSoC)*, June 29 to July 1, Bremen, Germany.

Palma, J.C., de Mello, A.V., Moller, L., Moraes, F., and Calazans, N. 2002. Core communication interface for FPGAs. In *Proceedings of the 15th Annual Symposium on Integrated Circuits and Systems Design*, September 9–14, Porto Alegre, Brazil.

Pionteck, T., Koch, R., and Albrecht, C. 2006. Applying partial reconfiguration to networks-on-chips. In *Proceedings of the 16th International Conference on Field-Programmable Logic and Applications*, August 28–30, Madrid, Spain.

Rodriguez, A., Valverde, J., Castanares, C., Portilla, J., de la Torre, E., and Riesgo, T. 2015. Execution modeling in self-aware FPGA-based architectures for efficient resource management. In *Proceedings of the 10th International Symposium on Reconfigurable Communication-Centric Systems-on-Chip (ReCoSoC)*, June 29 to July 1, Bremen, Germany.

Salehie, M. and Tahvildari, L. 2009. Self-adaptive software: Landscape and research challenges. *ACM Transactions on Autonomous and Adaptive Systems*, 4(14):1–42.

Salvador, R., Otero, A., Mora, J., de la Torre, E., Riesgo, T., and Sekanina, L. 2013. Self-reconfigurable evolvable hardware system for adaptive image processing. *IEEE Transactions on Computers*, 62:1481–1493.

Santambrogio, M.D. 2009. From reconfigurable architectures to self-adaptive autonomic systems. In *Proceedings of the International Conference on Computational Science and Engineering (CSE 2009)*, August 29–31, Vancouver, British Columbia, Canada.

Sohanghpurwala, A.A., Athanas, P., Frangieh, T., and Wood, A. 2011. OpenPR: An open-source partial reconfiguration toolkit for Xilinx FPGAs. In *Proceedings of the 2011 IEEE International Symposium on Parallel and Distributed Processing Workshops and Phd Forum (IPDPSW 2011)*, May 16–20, Anchorage, AK.

Steiner, N. and Athanas, P. 2009. Hardware autonomy and space systems. In *Proceedings of the 2009 IEEE Aerospace Conference*, March 7–14, Big Sky, MT.

Steiner, N., Wood, A., Shojaei, H., Couch, J., Athanas, P., and French, M. 2011. Torc: Towards an open-source tool flow. In *Proceedings of the 19th ACM/SIGDA International Symposium on Field Programmable Gate Arrays (FPGA '11)*, February 27 to March 1, Monterey, CA.

Ullmann, M., Hubner, M., Grimm, B., and Becker, J. 2004. An FPGA run-time system for dynamical on demand reconfiguration. In *Proceedings of the 18th International Parallel and Distributed Processing Symposium*, April 26–30, Santa Fe, NM.

Valverde, J., Rodriguez, A., Mora, J., Castañares, C., Portilla, J., de la Torre, E., and Riesgo, T. 2014. A dynamically adaptable image processing application trading off between high performance, consumption and dependability in real time. In *Proceedings of the 2014 International Conference on Field Programmable Logic and Applications (FPL 2010)*, August 31 to September 2, Munich, Germany.

Walder, H. and Platzner, M. 2004. A runtime environment for reconfigurable hardware operating systems. In *Proceedings of the 14th International Conference on Field-Programmable Logic and Applications*, August 30 to September 1, Leuven, Belgium.

9

Industrial Electronics
Applications of FPGAs

9.1 Introduction

From all that has been discussed in previous chapters, it is evident that FPGA vendors are continuously devoting efforts to include in their devices new features, or improvements to existing ones (as well as in their design tools), aiming at an increasing penetration in the digital design market. Today, FPGAs are used in many different industrial applications because of their high speed and flexibility, inherent parallelism, good cost–performance trade-off, and huge variety of available specialized logic resources. As a consequence, they have been extensively analyzed over the years from the perspective of industrial electronics (Monmasson and Cirstea 2007, 2013; Naouar et al. 2007; Rodriguez-Andina et al. 2007, 2015; Monmasson et al. 2011a,b; Gomes et al. 2013; Gomes and Rodriguez-Andina 2013).

Thanks mainly to the paradigm shift enabled by the availability of increasingly powerful, feature-rich, heterogeneous FPSoCs (Figure 9.1), they are expected to consolidate their application domains and enter new ones. In the authors' view, although microcontrollers will keep being the dominating solution in the market (because they are the best choice for many low- to mid-complexity embedded systems), the role of FPGAs will be increasingly significant both for the implementation of high-performance peripherals and in high-complexity systems.

One of the main impediments for a wider use of FPGAs in the industry has been the limited knowledge of the technology by nonspecialists in hardware design. It is our hope that this book contributes to mitigate this problem by showing the advantages that can be taken from the most current FPGA devices in industrial applications in an accessible way for these nonspecialists.

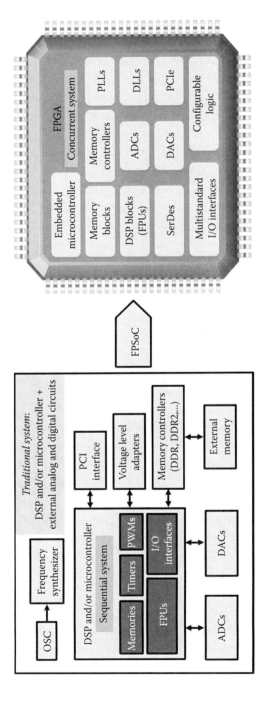

FIGURE 9.1

The FPSoC paradigm shift in digital systems design.

A comprehensive analysis of all current and prospective industrial applications of FPGAs would be extremely lengthy. Therefore, to conclude this book, three main design areas (advanced control techniques, electronic instrumentation, and digital real-time simulation) and three very significant industrial application domains (mechatronics, robotics, and power systems design) are concisely addressed in this chapter. This choice is based on the fact that, in these application domains, there is a current demand for systems capable of providing fast digital implementations of complex algorithms, for example, for switching purposes.

9.2 FPGA Application Domains in Industrial Electronics

Although (relatively) simple, FPGAs are still being used for glue logic implementation or rapid prototyping purposes (which were their original usages), the ever-increasing variety and amount of hardware resources, as well as the ability of the most current devices to implement whole complex systems in a single chip, has tremendously widened their application domains. The original applications have also benefited from this evolution so that, for instance, HIL approaches can be implemented instead of just basic system prototyping. Moreover, as highlighted throughout the previous chapters, current FPGAs are particularly suitable for configurable computing applications and for developing dynamically reconfigurable systems. In spite of all this, many complex digital systems are still today built from the combination of separate digital processing and configurable logic chips, such as in Hasanzadeh et al. (2014) or Wen et al. (2014). Very interestingly, migration from these platforms to FPSoC-based ones is relatively easy because, for instance, the existing software can be reused in embedded processors.

In the following sections, the aforementioned significant industrial applications of FPGAs are briefly discussed.

9.2.1 Digital Real-Time Simulation of Power Systems

Hardware-in-the-loop simulation (HILS) is very useful, among other applications, for real-time simulation of power systems, where it allows development time, cost, and time to market to be reduced (Buccella et al. 2012). In addition, it allows these systems to be safely checked under faulty (Zhu et al. 2014) or extreme operating conditions. A fundamental requirement of HILS in this context is to achieve very short time steps, which can be provided by FPGAs. This, together with the availability of memory and complex arithmetic blocks, as well as their parallelism and reconfigurability, makes FPGAs very suitable—and increasingly used—HILS platforms.

Many existing HILS solutions are based on the use of DSPs as main processing elements and FPGAs as specialized coprocessors and/or communication links (Wang et al. 2013; Hasanzadeh et al. 2014; Schmitt et al. 2014). Interfaces with analog inputs and outputs are implemented with separate ADC and DAC chips. It is worth noting that faster and more compact solutions can be achieved by migrating these structures to current FPSoC platforms, where most of the required resources are embedded. For instance, an FPGA-only platform is reported in Dagbagi et al. (2013), where a PWM rectifier is emulated in real time in one device and the corresponding controller is implemented in another one as an FPSoC, using an ARM Cortex-M3 core.

9.2.2 Advanced Control Techniques

9.2.2.1 Power Systems

Digital control of power systems has been gaining increasing interest over the years. Current solutions, usually based on microcontrollers and DSPs, suffer from relatively long execution times and limited flexibility to interface with analog signals. These issues can be successfully addressed with modern FPGAs. Actually, the analysis in Monmasson et al. (2011b) concludes that hardware solutions are the only viable options in practice for systems with strong timing requirements. Very interestingly, some of the most challenging needs identified in that analysis—efficient floating-point computations and analog/digital interfaces—can be addressed with the specialized hardware resources available in current FPGAs.

Many designers have already realized that FPGAs are the best solution to generate control signals for switching devices in power converters (Hwang et al. 2013; Lu et al. 2013; Morales-Caporal et al. 2013; Miura et al. 2014; Wang et al. 2014), particularly when the number of switches, the switching frequency, or both are high, which is the case in many modern converter topologies. However, in most of these works, there remains an inertia to implement the main control loops (and other fundamental tasks, such as filtering) in separate microcontroller or DSP chips and to use external ADCs, instead of taking advantage of FPSoC architectures. For instance, in Senicar et al. (2014), an FPGA is used to implement a fast current control loop in hardware, whereas higher-level control loops—speed and position—are implemented in a separate microcontroller chip, and external ADCs are used. Similarly, in Prabhala et al. (2012), the current controller for a voltage source converter is implemented in an FPGA, and the grid synchronization PLL in a DSP. The problems identified in Wen et al. (2014), related to the limited speed and processing power of DSPs, could be alleviated by using FPGAs to concurrently run some part of the proposed computation algorithm. Other reported systems that could have benefited from the use of FPSoCs are those of Kobravi et al. (2013) and Smidl et al. (2013).

Also, in the area of power systems, fault tolerance is a fundamental requirement when continuous operation must be ensured. In this case, the faster the fault detection, the safer the operation of the system. The usefulness of FPGAs for this purpose is demonstrated in Shahbazi et al. (2013).

In general, digital control of power systems is an area where the use of FPSoCs should be expected to grow significantly in the near future.

9.2.2.2 Robotics and Automotive Electronics

High operating frequency and low latency are fundamental requirements for control loops in many robotic applications, where at the same time complex algorithms are also needed for various purposes. Although DSPs—sometimes combined with FPGAs—are extensively used in this domain, their sequential nature in many cases may result in excessive latency when executing those algorithms. In addition, DSPs provide limited flexibility to include additional sensors or actuators, in turn limiting system scalability. In contrast, current FPGA devices offer better performance and accuracy, allowing real-time or fault-tolerant operation to be achieved (Hace and Franc 2013; May and Krougjicof 2013; Phuong et al. 2014).

Modular robotics (e.g., for humanoid robots) can take advantage of current FPGA devices to achieve higher sampling frequencies—and, then, better performance—than centralized systems. The flexibility of FPSoC-based solutions adds to that of these modular systems, allowing robot structures to be easily modified, extended, or repaired, even in the field. In this context, the availability of FPGAs including powerful embedded processors contributes to reduced size and weight, as well as lower processor intercommunication latency. Examples of current use of FPGAs in this area can be found in Zhu et al. (2013) and Pierce and Cheng (2014).

The application of FPGAs in automotive systems is an area of increasing interest, among other factors, because of the need to perform very complex operations with the low latency required by safety-related functions, such as the antilock braking system or the electronic stability program (Guo et al. 2013). Similar problems are addressed in Lu et al. (2015) for controlling dual-clutch transmission gearshifts. The increasingly powerful embedded processors and DSP blocks available in current FPGAs should play a key role in a wider penetration of these devices into automotive applications.

9.2.2.3 Use of Floating-Point Operations

There is an increasing number of applications requiring the use of floating-point operations to achieve the target performance (Guo et al. 2013; Hwang et al. 2013; Sepulveda et al. 2013; Barranco et al. 2014). Although efficient resources for floating-point operations were traditionally not available in FPGAs, it is currently possible to implement them either in embedded processors (Jimenez et al. 2014) or in dedicated specialized hardware blocks (Juarez-Abad et al. 2014; Liu and Dinavahi 2014). Even better results could be achieved by using the IEEE 754–compatible DSP blocks available in the most current devices.

9.2.3 Electronic Instrumentation

Same as floating-point operators, the lack of embedded ADCs and DACs was one of the main traditional limitations in FPGAs. The availability of these resources in recent devices (as well as of others, such as complex PLLs) opens the door for the improvement of many industrial systems that could not benefit from them when they were developed, such as the power systems presented in Hwang et al. (2013) and Guzinski and Abu-Rub (2013) or the ones related to sensors and microelectromechanical systems reported in Cheng et al. (2013), Xia et al. (2013), and Xu (2014, 2015).

An area where FPGAs provide particularly good performance at a reasonable complexity/cost is that of vision systems, in applications such as intelligent spaces (Rodriguez-Araujo et al. 2014) or unmanned vehicles (May and Krougjicof 2013). For instance, the configurable hardware system for traffic sign recognition, presented in Aguirre-Dobernack et al. (2013), achieves 60 fps with 1280 × 720 pixel images.

In aerial unmanned vehicles, huge amounts of data must be processed with low latency for purposes such as flight control, real-time mapping, or obstacle avoidance. Reduced size and weight are also fundamental requirements of these systems, which are more easily fulfilled with single-chip FPSoC implementations than with multichip heterogeneous solutions, such as the one presented in Schmid and Hirschmuller (2013), consisting of two cameras, an Intel Core2Duo processor board, a Spartan 6 FPGA board, an ARM processor board, and an IMU. By using an FPGA including ARM Cortex-A9 processors, the number of required boards—and thus the size and weight of the system—could be greatly reduced. The system presented in Nikolic et al. (2014) for real-time simultaneous localization and mapping consists of a Zynq-7000 FPSoC, four cameras, and an IMU, where the Zynq-7000 device replaces the combination of an FPGA and an Intel ATOM processor used in previous prototypes.

Another application where an advantage can be taken from the FPGA implementation of image processing algorithms is industrial laser cladding. In the work by Rodriguez-Araujo et al. (2012), an adaptive fuzzy PI controller for laser cladding systems is proposed, which works with data extracted from a real-time image processing system (achieving 100 Hz sampling rate for 800 × 600 pixel images). The whole monitoring and control system is implemented in an FPGA, combining a Nios II soft processor with high-performance custom hardware peripherals.

9.3 Conclusion

Taking advantage of the latest advancements in microelectronics fabrication technologies, FPGA vendors are continuously looking for new features to be included in their devices, aiming at making them suitable for wider application areas. In some cases, they are just extensions or improvements to previously available resources, whereas in other cases, architectures are extended with new resources aimed at mitigating traditional limitations (e.g., for interfacing with analog signals or for efficient computation of floating-point operations) or at improving performance (e.g., regarding memory controllers or serial communication interfaces).

Thanks to all these advancements, FPGAs cannot be considered just configurable hardware devices anymore, but they have to be approached with the FPSoC design paradigm in mind. Although in the authors' opinion there is still much room for improvements, vendors and third parties have also to be acknowledged for their continuous efforts in improving design tools, particularly considering the high-level approach required by FPSoCs, as well as the need for making the technology readily accessible to application engineers, who, in their vast majority, are not hardware designers.

The advantages of FPSoCs have hopefully been shown throughout this book to be many and significant. This is particularly important to realize for designers working in specific applications, who are (obviously) mainly focused on the details of their target applications rather than on implementation details, so they tend to keep basing their designs on the same devices they are used to working with. A similar situation occurred when digital control of power converters became a—very advantageous—practical possibility, but most designers continued to use for years the analog solutions they were familiar with and knew how to tune for their purposes. The current challenge for designers in the FPSoC area is to have enough knowledge in both the hardware and software domains, their integration, and the partitioning of tasks between them.

References

Aguirre-Dobernack, N., Guzman-Miranda, H., and Aguirre, M.A. 2013. Implementation of a machine vision system for real-time traffic sign recognition on FPGA. In *Proceedings of the 39th Annual Conference of the IEEE Industrial Electronics Society (IECON'2013)*, November 10–13, Vienna, Austria.

Barranco, F., Diaz, J., Pino, B., and Ros, E. 2014. Real-time visual saliency architecture for FPGA with top-down attention modulation. *IEEE Transactions on Industrial Informatics*, 10(3):1726–1735.

Buccella, C., Cecati, C., and Latafat, H. 2012. Digital control of power converters—A survey. *IEEE Transactions on Industrial Informatics*, 8(3):437–447.

Cheng, Ch.-F., Li, R.-S., and Chen, J.-R. 2013. Design of the DC leakage current sensor with magnetic modulation-based scheme. In *Proceedings 2013 IEEE International Symposium on Industrial Electronics (ISIE'2013)*, May 28–31, Taipei, Taiwan.

Dagbagi, M., Hemdani, A., Idkhajine, L., Naouar, M.W., Monmasson, E., and Slama-Belkhodja, I. 2013. FPGA-based real-time hardware-in-the-loop validation of a 3-phase PWM rectifier controller. In *Proceedings of the 39th Annual Conference of the IEEE Industrial Electronics Society (IECON'2013)*, November 10–13, Vienna, Austria.

Gomes, L., Monmasson, E., Cirstea, M., and Rodriguez-Andina, J.J. 2013. Industrial electronic control: FPGAs and embedded systems solutions. In *Proceedings of the 39th Annual Conference of the IEEE Industrial Electronics Society (IECON'2013)*, November 10–13, Vienna, Austria.

Gomes, L. and Rodríguez-Andina, J.J. 2013. Guest editorial special section on embedded and reconfigurable systems. *IEEE Transactions on Industrial Informatics*, 9(3):1588–1590.

Guo, H., Chen, H., Xu, F., Wang, F., and Lu, G. 2013. Implementation of EKF for vehicle velocities estimation on FPGA. *IEEE Transactions on Industrial Electronics*, 60(9):3823–3835.

Guzinski, J. and Abu-Rub, H. 2013. Speed sensorless induction motor drive with predictive current controller. *IEEE Transactions on Industrial Electronics*, 60(2):699–709.

Hace, A. and Franc, M. 2013. FPGA implementation of sliding-mode-control algorithm for scaled bilateral teleoperation. *IEEE Transactions on Industrial Informatics*, 9(3):1291–1300.

Hasanzadeh, A., Edrington, C.S., Stroupe, N., and Bevis, T. 2014. Real-time emulation of a high-speed microturbine permanent-magnet synchronous generator using multiplatform hardware-in-the-loop realization. *IEEE Transactions on Industrial Electronics*, 61(6):3109–3118.

Hwang, S.-H., Liu, X., Kim, J.-M., and Li, H. 2013. Distributed digital control of modular-based solid-state transformer using DSP+FPGA. *IEEE Transactions on Industrial Electronics*, 60(2):670–680.

Jimenez, O., Lucia, O., Urriza, I., Barragan, L.A., and Navarro, D. 2014. Analysis and implementation of FPGA-based online parametric identification algorithms for resonant power converters. *IEEE Transactions on Industrial Informatics*, 10(2):1144–1153.

Juarez-Abad, J.A., Linares-Flores, J., Guzmán-Ramirez, E., and Sira-Ramirez, H. 2014. Generalized proportional integral tracking controller for a single-phase multilevel cascade inverter: An FPGA implementation. *IEEE Transactions on Industrial Informatics*, 10(1):256–266.

Kobravi, K., Iravani, R., and Kojori, H.A. 2013. Three-leg/four-leg matrix converter generalized modulation strategy—Part II: Implementation and verification. *IEEE Transactions on Industrial Electronics*, 60(3):860–872.

Liu, J. and Dinavahi, V. 2014. A real-time nonlinear hysteretic power transformer transient model on FPGA. *IEEE Transactions on Industrial Electronics*, 61(7):3587–3597.

Lu, X., Chen, H., Gao, B., Zhang, Z., and Jin, W. 2015. Data-driven predictive gearshift control for dual-clutch transmissions and FPGA implementation. *IEEE Transactions on Industrial Electronics*, 62(1):599–610.

Lu, Z.-G., Zhao, L.-L., Zhu, W.-P., Wu, C.-J., and Qin, Y.-S. 2013. Research on cascaded three-phase-bridge multilevel converter based on CPS-PWM. *IET Power Electronics*, 6(6):1088–1099.

May, K. and Krougjicof, N.K. 2013. Moving target detection for sense and avoid using regional phase correlation. In *Proceedings 2013 IEEE International Conference on Robotics and Automation (ICRA'2013)*, May 6–10, Karlsruhe, Germany.

Miura, Y., Inubushi, K., Ito, M., and Ise, T. 2014. Multilevel modular matrix converter for high voltage applications. Control, design and experimental characteristics. In *Proceedings of the 40th Annual Conference of the IEEE Industrial Electronics Society (IECON'2014)*, October 30 to November 1, Dallas, TX.

Monmasson, E. and Cirstea, M. 2013. Guest editorial special section on industrial control applications of FPGAs. *IEEE Transactions on Industrial Informatics*, 9(3):1250–1252.

Monmasson, E. and Cirstea, M.N. 2007. FPGA design methodology for industrial control systems—A review. *IEEE Transactions on Industrial Electronics*, 54(4):1824–1842.

Monmasson, E., Idkhajine, L., Cirstea, M.N., Bahri, I., Tisan, A., and Naouar, M.W. 2011a. FPGAs in industrial control applications. *IEEE Transactions on Industrial Informatics*, 7(2):224–243.

Monmasson, E., Idkhajine, L., and Naouar, M.W. 2011b. FPGA-based controllers. *IEEE Industrial Electronics Magazine*, 5(1):14–26.

Morales-Caporal, R., Bonilla-Huerta, E., Hernández, C., Arjona, M.A., and Pacas, M. 2013. Transducerless acquisition of the rotor position for predictive torque controlled PM synchronous machines based on a DSP-FPGA digital system. *IEEE Transactions on Industrial Informatics*, 9(2):799–807.

Naouar, M.-W., Monmasson, E., Naassani, A.A., Slama-Belkhodja, I., and Patin, N. 2007. FPGA-based current controllers for AC machine drives—A review. *IEEE Transactions on Industrial Electronics*, 54(4):1907–1925.

Nikolic, J., Rehder, J., Burri, M., Gohl, P., Leutenegger, S., Furgale, P.T., and Siegwart, R. 2014. A synchronized visual-inertial sensor system with FPGA preprocessing for accurate real-time SLAM. In *Proceedings 2014 IEEE International Conference on Robotics and Automation (ICRA'2014)*, May 31 to June 7, Hong Kong, China.

Phuong, T.T., Ohishi, K., Yokokura, Y., and Mitsantisuk, C. 2014. FPGA-based high-performance force control system with friction-free and noise-free force observation. *IEEE Transactions on Industrial Electronics*, 61(2):994–1008.

Pierce, B. and Cheng, G. 2014. Versatile modular electronics for rapid design and development of humanoid robotic subsystems. In *Proceedings 2014 IEEE/ASME International Conference on Advanced Intelligent Mechatronics (AIM'2014)*, July 8–11, Besancon, France.

Prabhala, V.A., Cespedes, M., and Sun, J. 2012. Implementation of DQ domain control in DSP and FPGA. In *Proceedings of the 2012 Twenty-Seventh Annual IEEE Applied Power Electronics Conference and Exposition (APEC'2012)*, February 5–9, Orlando, FL.

Rodriguez-Andina, J.J., Moure, M.J., and Valdes, M.D. 2007. Features, design tools, and application domains of FPGAs. *IEEE Transactions on Industrial Electronics*, 54(4):1810–1823.

Rodriguez-Andina, J.J., Valdes, M.D., and Moure, M.J. 2015. Advanced features and industrial applications of FPGAs—A review. *IEEE Transactions on Industrial Informatics*, 11(4):853–864.

Rodriguez-Araujo, J., Rodriguez-Andina, J.J., Farina, J., and Chow, M.-Y. 2014. Field-programmable System-on-Chip for localization of UGVs in an indoor iSpace. *IEEE Transactions on Industrial Informatics*, 10(2):1033–1043.

Rodriguez-Araujo, J., Rodriguez-Andina, J.J., Farina, J., Vidal, F., Mato, J.L., and Montealegre, M.A. 2012. Industrial laser cladding systems: FPGA-based adaptive control. *IEEE Industrial Electronics Magazine*, 6(4):35–46.

Schmid, K. and Hirschmuller, H. 2013. Stereo vision and IMU based real-time egomotion and depth image computation on a handheld device. In *Proceedings 2013 IEEE International Conference on Robotics and Automation (ICRA'2013)*, May 6–10, Karlsruhe, Germany.

Schmitt, A., Richter, J., Jurkewitz, U., and Braun, M. 2014. FPGA-based real-time simulation of nonlinear permanent magnet synchronous machines for power hardware-in-the-loop emulation systems. In *Proceedings of the 40th Annual Conference of the IEEE Industrial Electronics Society (IECON'2014)*, October 30 to November 1, Dallas, TX.

Senicar, F., Dopker, M., Bartsch, A., Kruger, B., and Soter, S. 2014. Inverter based method for measurement of PMSM machine parameters based on the elimination of power stage characteristics. In *Proceedings of the 40th Annual Conference of the IEEE Industrial Electronics Society (IECON'2014)*, October 30 to November 1, Dallas, TX.

Sepulveda, C.A., Munoz, J.A., Espinoza, J.R., Figueroa, M.E., and Baier, C.R. 2013. FPGA v/s DSP performance comparison for a VSC-based STATCOM control application. *IEEE Transactions on Industrial Informatics*, 9(3):1351–1360.

Shahbazi, M., Poure, P., Saadate, S., and Zolghadri, M.R. 2013. FPGA-based reconfigurable control for fault-tolerant back-to-back converter without redundancy. *IEEE Transactions on Industrial Electronics*, 60(8):3360–3371.

Smidl, V., Nevdev, R., Kosan, T., and Peroutka, Z. 2013. FPGA implementation of marginalized particle filter for sensorless control of PMSM drives. In *Proceedings of the 39th Annual Conference of the IEEE Industrial Electronics Society (IECON'2013)*, November 10–13, Vienna, Austria.

Wang, C., Li, W., and Belanger, J. 2013. Real-time and faster-than-real-time simulation of modular multilevel converters using standard multi-core CPU and FPGA chips. In *Proceedings of the 39th Annual Conference of the IEEE Industrial Electronics Society (IECON'2013)*, November 10–13, Vienna, Austria.

Wang, Y., Yang, K., He, C., and Chen, G. 2014. A harmonic elimination approach based on moving average filter for cascaded DSTATCOM. In *Proceedings of the 40th Annual Conference of the IEEE Industrial Electronics Society (IECON'2014)*, October 30 to November 1, Dallas, TX.

Wen, H., Cheng, D., Teng, Z., Guo, S., and Li, F. 2014. Approximate algorithm for fast calculating voltage unbalance factor of three-phase power system. *IEEE Transactions on Industrial Informatics*, 10(3):1799–1805.

Xia, G.-M., Qiu, A.-P., Shi, Q., and Yan, S. 2013. Test and evaluation of a silicon resonant accelerometer implemented in SOI technology. In *Proceedings 2013 IEEE Sensors*, November 3–6, Baltimore, MD.

Xu, Q. 2014. Design and smooth position/force switching control of a miniature gripper for automated microhandling. *IEEE Transactions on Industrial Informatics*, 10(2):1023–1032.

Xu, Q. 2015. Robust impedance control of a compliant microgripper for high-speed position/force regulation. *IEEE Transactions on Industrial Electronics*, 62(2):1201–1209.

Zhu, W., Lamarche, T., Dupuis, E., Jameux, D., Barnard, P., and Liu, G. 2013. Precision control of modular robot manipulators: The VDC approach with embedded FPGA. *IEEE Transactions on Robotics*, 29(5):1162–1179.

Zhu, Z., Li, X., Rao, H., Wang, W., and Li, W. 2014. Testing a complete control and protection system for multi-terminal MMC HVDC links using hardware-in-the-loop simulation. In *Proceedings of the 40th Annual Conference of the IEEE Industrial Electronics Society (IECON'2014)*, October 30 to November 1, Dallas, TX.

Index